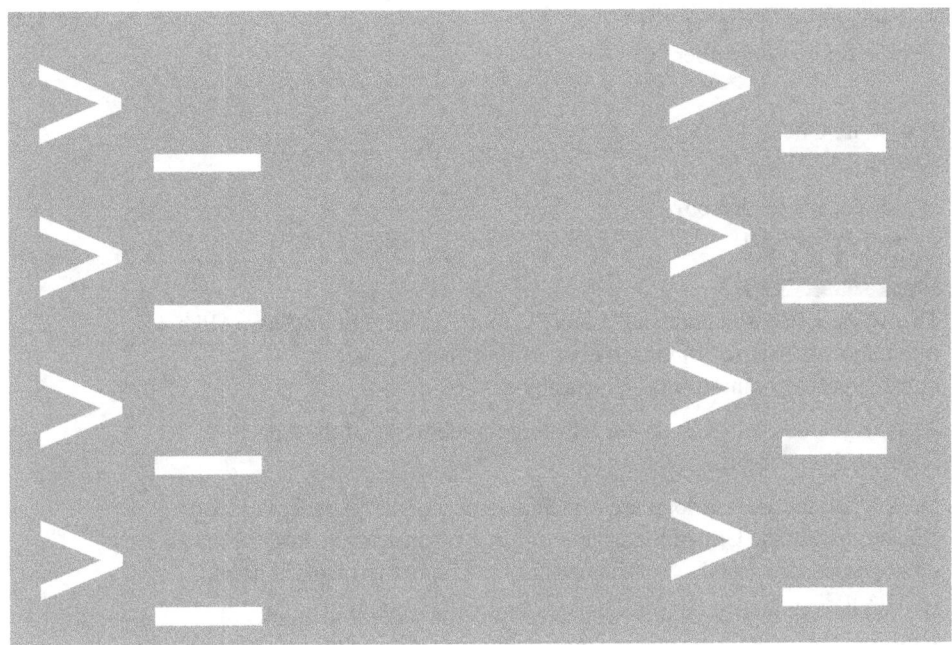

DOCS LIKE CODE

Collaborate and Automate
to Improve Technical Documentation

BY ANNE GENTLE

Third Edition

Copyright © 2023 Anne Gentle

ISBN 978-1-387-53149-3

First edition: May 2017
Second edition: October 2017
Third edition: November 2022

Just Write Click
Austin, TX, USA

https://docslikecode.com

This work is licensed under a Creative Commons Attribution-ShareAlike 4.0 International License. To view a copy of this license, visit https://creativecommons.org/licenses/by-sa/4.0/.

Apache, jclouds, and Maven are registered trademarks of the Apache Software Foundation.

Docker and the Docker logo are trademarks or registered trademarks of Docker, Inc. in the United States and/or other countries. Docker, Inc. and other parties may also have trademark rights in other terms used herein.

Git and the Git logo are either registered trademarks or trademarks of Software Freedom Conservancy, Inc., corporate home of the Git Project, in the United States and/or other countries.

GitHub is a trademark registered in the United States by GitHub, Inc.

Linux is the registered trademark of Linus Torvalds in the U.S. and other countries.

The OpenStack Word Mark and OpenStack Logo are either registered trademarks/service marks or trademarks/service marks of the OpenStack Foundation, in the United States and other countries and are used with the OpenStack Foundation's permission. We are not affiliated with, endorsed or sponsored by the OpenStack Foundation, or the OpenStack community.

Oracle and Java are registered trademarks of Oracle and/or its affiliates.

Python is a registered trademark of the Python Software Foundation.

Rackspace is a registered trademark of Rackspace US, Inc.

Red Hat and OpenShift are trademarks of Red Hat, Inc., registered in the U.S. and other countries.

Any use of a trademarked name without a trademark symbol is for readability purposes only. We have no intention of infringing on the trademark.

Disclaimer

The information in this book is provided without warranty. The author, contributors, and publisher have neither liability nor responsibility to any person or entity related to any loss or damages arising from the information contained in this book.

Contents

Foreword

Docs as code has changed a lot since the first edition of this book was written. It has now become an accepted way to integrate development and documentation teams and produce high-quality documentation for software products. The docs as code approach has become a more common way for software organizations to approach creating and shipping documentation projects.

Throughout my career as a developer, I've seen the wonders that come from getting documentation and code in the same workflow. My first experience with this was co-founding *Read the Docs*[1], which is a service that builds and hosts software documentation. Our audience is primarily developers, and by bringing docs inside the development workflow, developers have embraced documentation at a massive scale. For us, massive scale means growing from 28,000 projects in 2016 to more than 80,000 projects in 2022 and serving over 55 million pages of documentation a month, close to being one of the top-1000 sites on the internet.

While *Read the Docs* has been successful at getting open source developers to write documentation, it felt like documentation still wasn't highly valued. So in 2013, I helped co-found *Write the Docs*[2], which advocates for the value of documentation in the software industry. The community there has taught me many things, one of the biggest being that technical writers get more collaboration and ownership from their product teams in a docs as code workflow.

---

[1]   https://readthedocs.org
[2]   https://www.writethedocs.org/

Starting with a *talk from Google*[3] in 2015 about how they have adopted docs as code, the approach has spread across the entire software ecosystem. We have talks from each year since in our *Docs as Code*[4] page, which you can visit for more information about this approach. The docs as code approach has scaled up within technical organizations like Google, Twitter, and Microsoft as well as government and non-profit organizations. It has also scaled down to single-person open source projects, and it's used heavily by many businesses and projects in between.

This book is a practical introduction to the docs as code mindset. It gives you an overview of the tools and processes that allow you to integrate a docs as code approach into your team. From making small adjustments in workflows, to transforming how your team works, you will get value out of the information in this book.

Anne, Diane, and Kelly have worked on some of the largest open source code bases, have gone through multiple documentation transformations, and have been at multiple companies observing docs as code in many contexts. They distill their experiences into actionable advice, along with case studies that show how it works in practice. The book includes hard-won advice that comes from years of putting these ideas into practice.

The docs as code concepts continue to evolve, and this book is the best introduction to docs as code that exists. I hope that you enjoy reading it, and are able to introduce the practices described into your work. I truly believe that treating docs like code is the future of software documentation, and I'm glad you're along for the ride :)

Eric Holscher

Cofounder of Read the Docs & Write the Docs
Portland, Oregon
Jan 30, 2017

---

[3] https://www.youtube.com/watch?v=EnB8GtPuauw
[4] https://www.writethedocs.org/guide/docs-as-code/#docs-as-code-at-write-the-docs

About this book

> Authors

> How this book was created

The authors of this book are academically-trained old-school technical writers who targeted that same audience in the first version of this book.

Although technical writers and their managers are the intended audiences, this edition also welcomes developers who want to implement a docs-as-code system.

This revised edition targets technical writing teams that want to either:

- Move to a docs-as-code system.
- Improve an existing docs-as-code system.

Feedback on the first and second editions indicates that some technical writers think that docs-as-code workflows are the only choice in software-development environments. This book clarifies that docs-as-code frameworks are not always the ideal choice and helps readers determine when to implement a docs-as-code system.

Authors

Anne Gentle

Anne Gentle is a director for developer experience at Cisco for the DevNet community, a developer relations program for Cisco programmable platforms. With her team of experts, she supports developer tools for API design, developer documentation, and developer education including infrastructure integration. At Cisco, her team enables more than a thousand contributors to use a docs-as-code system on Enterprise GitHub to publish to developer.cisco.com[1]. She wrote this book, *Docs Like Code* to demonstrate developer tools and workflows like GitHub and automated publishing and code integration, which are applied in the technical writing world.

She serves on the Workforce Advisory Committee at Austin Community College, advancing the field into API and developer documentation. In collaboration with the Social Justice Action Office at Cisco, Anne works with engineering teams to eliminate biased language in code and content. From 2010-2016, Anne led hundreds of OpenStack contributors in growing the API docs from a few APIs to many APIs.

Anne writes about various topics at *Just Write Click*[2] and shares more dynamic content than the book on the *Docs as Code*[3] website. You are welcome to share your stories on that site as well.

Diane Skwish

Diane Skwish currently works as a Senior Technical Writer - API Documentation at Indeed in Austin, Texas. She learned GitHub when she worked at Rackspace on REST API docs for both Rackspace products and OpenStack services and later as one of the main REST API docs writers at PayPal. At first, the switch from a traditional, pessimistic locking system to an optimistic GitHub system was challenging. She appreciates the docs-as-code

---

[1] https://developer.cisco.com
[2] https://justwriteclick.com
[3] https://docsascode.com

community where everyone is encouraged to contribute content rather than own books.

Contributor Kelly Holcomb

Kelly Holcomb is a Principal Technical Editor for Oracle Cloud Infrastructure, where she customizes Acrolinx for authors to edit using the rules and style guidance she created for the OCI organization. Previously at Rackspace, she worked with multiple writing and development teams using GitHub for docs. Although she's been correcting people's grammar since middle school, high-tech companies such as Oracle, Rackspace, and BMC Software have been paying her to do it for more than 20 years. As a technical editor, she loves to take any piece of writing and make it better. And although she shed a few, hot tears while learning GitHub, she's growing to appreciate the collaborative environment that it provides.

How this book was created

Anne, Diane, and Kelly worked at Rackspace in 2015. Each contributed their expertise to transform how information was written, produced, and edited.

With Anne as the chief cook and bottle-washer (and instigator), they wrote this book together to chronicle the docs-as-code way of working. They maintained the book's source files in a private GitHub repo, authored the source files in Markdown, and configured the repo to use *GitBook*[4] for automatic builds of PDF, EPUB, and HTML output. For the third edition, the builds switched to *Bookdown*[5].

Diane and Anne wrote furiously for a few months, borrowing from their experiences of using and teaching others to use these techniques.

Kelly first edited the chapters individually by creating Word documents from the Markdown files, marking them up, and

[4] https://gitbook.com
[5] https://bookdown.org/

creating PDFs from them to make the tracked changes easier to read. Then, she created issues for each chapter's edits, and Diane and Anne made pull requests with the suggested edits. As a final quality check, Kelly edited the book as a whole, creating pull requests with final copy edits.

For the third edition, Anne updated content in Markdown, sending pull requests to Diane for edits, reviews, and approvals.

Drawing on equal parts inspiration and perspiration, the authors hope that this book introduces your team to docs-as-code techniques that improve their productivity and the docs.

Chapter 1

Introducing docs as code

> Why treat docs as code?

> Background for docs as code

To treat docs as code, you use developer tools and workflows. With the docs-as-code approach, contributors create and review source content in a Git-based repository, automatically run tests and build artifacts, and publish the artifacts without much human intervention.

For example, you can use Git, a distributed version control system, and GitHub or GitLab, which are collaborative coding and DevOps platforms, to create and maintain references, guides, tutorials, and other technical docs. You write, review, and deploy docs just as developers write, review, and deploy code.

Collaborative docs-as-code processes and tools work for both open-source and enterprise docs. Treating docs as code, you bring code and docs, and developers and writers, closer together. You dissolve any real or imagined boundaries between the two roles.

With docs as code, writing parallels coding.

You refactor content just as developers refactor code, by reorganizing and rewriting.

You review and edit another writer's docs just as developers review and update another developer's code. Rather than using sidebar comments in Word or Google Docs, you make inline comments in GitHub or GitLab.

You deploy websites, developer portals, PDFs, ePUBs, or web pages to target audiences just as developers deploy code.

You can use docs-as-code toolchains and workflows for all kinds of docs: open-source, enterprise, and personal docs. Docs-as-code toolchains and workflows have matured and become mainstream. You no longer find these techniques only in open source and developer documentation. Still, realize you may have considerations in your environment that would make docs as code a poor fit for your use case. Planning is essential, and people who collaborate and understand the workflow plus processes for them to use are equally necessary.

Although the title of this book is *Docs Like Code*, the current thinking is these are docs-as-code systems. For clarity, the rest of the book uses the phrase "docs as code" even when talking about earlier implementations of that system.

Docs as code can introduce more work, automate some work, and sometimes might not work at all. It's not a cure-all, automatic, nor will it fix all the difficulties with producing excellent technical documentation. It's complex and could cause more complexity as you push it forward or scale it across your organization.

But as you collaborate with teammates to apply docs-as-code techniques, you improve over time, and it can become your new way of working. Eventually, you find yourself saying, "I'm never moving away from my docs and code combinations."

Why treat docs as code?

Why and when do you want to treat docs as code? Which organizations can use these techniques? Could human resources or government organizations, for example, use them? What if contributors don't already have GitHub or GitLab accounts? Let's probe into the reasons to treat docs as code.

Defining *docs as code*

Over the years, docs have changed along with the technologies that they document. At their start, tech docs included formal specifications, assembly diagrams, and glossaries of technical terms. These docs targeted a literate audience with a manufacturing or mechanical background. Now, software docs are developed by using Agile techniques, and the docs themselves include interactive coding examples. But whether tech docs explain sewing machines or a REST API, their main goal has always been to aid in the effective use of technology.

When we say *docs*, we mean streamlined, tightly phrased, and fast-moving information that helps developers understand complex application interfaces. Docs can be anything from a single-page application to an entire developer reference site. Modern docs, with their web and mobile interfaces and supportive user experience, are purposeful, instructive, and even beautiful.

When we say *treat docs as code*, we mean that you:

- Store the doc source files in a version control system.
- Build the doc artifacts automatically.
- Ideally, run automated tests on the docs.
- Ensure that a trusted set of reviewers meticulously reviews the docs.
- Publish the artifacts without much human intervention.

These techniques constitute a docs-as-code framework.

Ideally, you seek ways to align the docs with the development systems that are already in place. You can use GitHub[1], Gerrit[2,] Bitbucket[3]®, Git[4], or GitLab[5], and other code systems to treat docs as code. This book leans heavily on experiences with Git, GitHub, and GitLab, but we do not intend to describe a one-size-fits-all solution here. We show how these tools provide a way to think

---

[1] https://github.com/
[2] https://www.gerritcodereview.com/
[3] https://bitbucket.org/
[4] https://git-scm.com/
[5] https://about.gitlab.com/

about including docs output with software and services, and how to leverage continuous integration and review systems while writing collaboratively.

What this book leaves out

The 2021 introduction of CoPilot by GitHub showed that Artificial Intelligence (AI) and Machine Learning (ML) systems can be taught to write code when given a large body of code from which to learn. As a natural extension of this training through code, companies provide tools that use AI/ML to write documentation. You use the tool to write code comments by highlighting the code in say, your Visual Studio Code editor, and asking the AI to write the documentation for that section of code.

The AI/ML documentation tool also seeks out and detects where documentation becomes outdated and notifies or sends alerts. This tool is a natural extension of Continuous Integration/Continuous Deployment with documentation. This type of code and docs integration is not what this book is about, though AI/ML opportunities are certainly the future of these techniques.

Reasons why teams treat docs as code

Alignment provides the best explanation for why teams move to docs as code. Tightly integrated development and docs teams likely share goals like performance, efficiency, and customer satisfaction. In a software world where Agile is often the way product teams produce software, the docs must be included in the definition of done. To get to that definition, it makes sense to use the same tools and build chains if these tools help with those shared goals.

Many forces, such as efficiency, profit, and user experience, drive organizations to integrate docs with code. Let's look at these forces:

Workplace evolution

If managers thought that workforces were starting distributed work before 2020, they had no idea how a global pandemic would fast-forward the need for remote, asynchronous work techniques. Everyone must communicate asynchronously with remote teammates since COVID19 sent many tech workers home from offices in March 2020. GitHub and code conventions provide standards that make distributed work easier and more efficient to do. Notifications provide collaboration advantages while letting individual contributors control their settings when they want to focus.

Technical teams are also convincing writers that they should join them on GitHub. Product managers and customer support representatives are coming to technical writing teams and saying, "I can't use your complex authoring tools, but I can write in Markdown. Is there a way for us to work together?"

For budgeting purposes, you might need to compare user licensing for Confluence, for example, which was US$5.50 per user in July 2022. If a developer or writer already has a GitHub user license, would a Confluence user license be considered an additional cost?

For some teams, docs-as-code is the only way to write. People on these teams have always used GitHub, Markdown, and static site generators to deliver websites for documentation.

Technology changes and opportunities

The rise of the REST API has been meteoric over the last two decades. The Programmable Web, an API directory, publishes its data for research purposes and showed that APIs show Faster Growth Rate in 2019 than Previous Years[6]. REST (REpresentational State Transfer) uses HTTP as the underlying protocol, which is "everywhere" as web services. REST systems provide an efficient and lightweight way to get data and perform actions on a web service. People use REST APIs to get weather data and social media posts, shop online, compare insurance quotes, and create cloud infrastructure, to

---

[6] https://www.mulesoft.com/infographics/api/rise-application-programming-interface

name only a few use cases. With the increasing use of software as a service (SaaS) and the need to document both application interfaces and user interfaces, excellent docs are in high demand.

GraphQL is another approach to web services APIs that provides the ability to collect, filter, and manipulate data. For data-intensive, distributed, and high-performance applications, GraphQL provides better data manipulation and querying than REST. GraphQL also requires developer documentation.

Webhooks, or ways to trigger an automatic action, have played a large part in the successful integration of GitHub with other tools such as instant messaging like Slack, build and test environments like Travis CI, and email. Because webhooks send an HTTP payload to a preconfigured URL, GitHub and GitLab can be used to do a multitude of tasks from builds to backups to deployments.

Market and business needs

Users expect docs to be delivered with the code. Developers expect to try tutorials with in-browser terminal windows. Plus you can find "try-it-out" buttons in the developer documentation for REST API services because interactive docs are the best way to showcase an API's capabilities.

Businesses are always looking for cost efficiencies, whether in tooling or headcount. The tooling for treating docs as code can be cost-effective, even when factoring in supportability. Leveraging the community is not a cost-saving measure necessarily, but it can provide otherwise unavailable resources. In some markets, senior developers make more than senior tech writers, but tech writers who can work in docs as code environments are earning high salaries due to their specialized knowledge and the efficiencies they create.

It's often the case that a tech writing team cannot document all the web services and APIs that your organization delivers. By distributing the work across development, test, and docs teams, you can provide documentation for more API endpoints. For example, imagine going from 30 API endpoints to 500 API endpoints. Your teams simply must find a way to work together on the docs and distribute both responsibility and accountability.

Goals of docs as code

When you treat docs as code, you accomplish the following goals:

Goal	Description
Distribute contributions	Collaborate with contributors efficiently by keeping docs in the same system as code with deliverables generated from source files.
Encourage technical contributions from subject matter experts	Enable experts who want to write tutorials or best practice guides in their favorite toolset.
Track doc bugs like code bugs	When you fix a doc bug, you reference that bug in the commit message to help reviewers judge whether the doc fix solves the stated problem.
Get better reviews	Trust team members to value docs, ensure technical accuracy and consistency, respect end users' needs, and advocate for the best doc deliverables for consumers.
Make useful docs	Front-end web development frameworks can make functional, modern docs or even web applications that work as docs.
Use developer tools, automation, and workflows	Apply software-development tools and techniques to software, API, and other technical docs. This set includes automated builds that let you and your teams focus on content.
Get value from cost-effective tools	Use pay-as-you-need, flexible subscriptions to purchase the right level of tool help. Also, authoring tools are often free or inexpensive per person.

Distribute contributions

By writing with and for your audience, you get to know your readers better. When you work with your readers on GitHub, you collaborate with your readers where they are.

One way that GitHub in particular promotes collaboration is through contribution graphs, which let writers and coders see who works in which areas. Contributors' reputations are laid out in green blocks in the graphs. You can use the information in the graphs to recognize and reward contributors and recruit the best talent. Contributors can use the graphs as an incentive to build a better online resume by GitHub metrics.

Treating docs as code is *not* about devaluing writing by outsourcing it to a ragtag bunch of contributors for free. GitHub Insights, for example, counts doc and dev contributions equally; visually, doc contributors are valued as highly as code and test contributors. Your collaborators have a shared purpose in making great docs. People often like to "pay it forward," reciprocating any support that they received, by writing helpful information for the docs.

GitHub and similar code systems avoid *documentation ghettos* because writers use the same tools that developers use. By adopting software techniques such as continuous integration, automated testing, and incremental improvement for docs, you get docs that are on par with the code itself. Also, you gain respect from those who value technical accuracy above all else, such as the app developers who consume your API docs. Keeping docs and code together also saves time and keeps context-switching to a minimum.

Sometimes, developers perceive that writers are distant from the product, and they consequently treat docs team members like second-class citizens. Written contributions, when treated like code, have the same value as developer code. We can show the metrics side-by-side.

In many open source projects, the patch and review processes are identical for code and docs, so the data used for event passes and voting privileges is the same for both developer contributions and technical writer contributions.

Lately, the trends have gone even further toward recognizing that it takes a wide variety of contributions to create successful projects. The GitHub docs team uses the All Contributors[7] project and speci-fication[8] to add contributions of all kinds. Answering questions, packaging, fundraising, organizing events, reporting bugs, testing (with users and in code), and reviewing, all of these tasks are recog-nized as contributions.

---

[7]   https://allcontributors.org/
[8]   https://allcontributors.org/docs/en/specification

Encourage technical contributions

Due to integrations and platform complexities, a single writer can't know an entire software product, nor can you hire a team to cover all integrations. For example, when GitHub Actions (an automation system for GitHub) was first introduced, the product went into the community while the writing team was still working on the docs. Out of the blue, a community member wrote a specific integration guide that a docs team member had not yet acquired the knowledge to write. In these scenarios, you must find specialists to collaborate with you. Those technical details can make a huge difference in the success of an information product, especially one targeted at developers or DevOps specialists. We recommend that you split deliverables into guides that encourage small but mighty contributions.

Your doc team cannot know everything about your product. Documenting integrations with other products is especially difficult. Outside contributors may know more than your team. Because many projects now use docs-as-code tools such as Git and Markdown, technical contributors can move between projects with known patterns.

Track doc bugs like code bugs

If you don't know what's wrong with your docs, you can't fix them. By treating docs as code and logging and triaging issues, you can make a plan to work through that long backlog of the broken parts of your docs and show incremental improvement over time.

Doc bugs can be tasks, outright errors, simple typos that make your docs look sloppy, or feature requests for the docs themselves. When you give your readers a way to track doc issues, they can tell you where they found confusing or incorrect content and even how to fix it.

Get better reviews

Review systems for enterprise docs can have workflows simply for mindless sign-offs, or for legal audits. When you treat docs as code, reviewers can see final changes in context before approving them.

Because code system workflows encourage small changes that are easy to compare, you can get more reviews. Also, you can encourage more interest in reviewing the docs by being in the notification system already used by developers on your team.

With automation, you can deploy a staging instance of the docs site so that reviewers and authors see the rendered version and not only the Markdown changes side-by-side.

Working in the same collaboration tools as technical people enables better reviews for technical content. Most GitHub repositories contain a file named `CONTRIBUTING.md` with instructions for both adding changes and reviewing those changes. Each reviewer must adhere to established review rules.

Make useful docs

By treating docs as code, you enter a community that truly wants interactive and useful docs sites. They want widely accepted tooling that is built in the open and shared by developers and writers alike.

Great docs themes are clean, useful, and helpful in organizing the docs. The Read the Docs theme is often referred to as a "gold standard." The theme is responsive for mobile devices, has an expandable table of contents, and has a best-in-class version implementation. The source code for the Sphinx theme[9] is available on GitHub.

You should spend time evaluating requirements for front-end web development and themes as part of your docs as code strategy and planning phase. Useful and functional docs have various requirements and lots of systems in the Git and Markdown toolchain can deliver them.

Use developer tools and workflows

Development teams perform continuous integration and deployment, completely mimicking their local environment in production and continuously updating the code and product. Docs can work the same way. Docs can be published automatically after trusted reviewers merge changes into the repository.

---

[9] https://github.com/readthedocs/sphinx_rtd_theme

Tests on the docs can ensure quality, such as white-space tests or spell checking. Docs can pass more complex tests as well, such as ensuring that API examples in the docs always work against the customer-facing web service. Docs can use continuous builds to ensure that they build and publish correctly so contributors can focus on content rather than on the build itself.

> **Tip**: Docs that can run an installation
>
> In the 2018 Europe Write the Docs conference, Predrag Mandic demonstrated the `rundoc` tool, a Python command-line utility from https://pypi.org/project/rundoc/ that runs fenced code blocks from within a Markdown file. The tool enabled him to document *and automate* an installation process that took two days to complete manually. He could use that tool to test any changes to the written doc as well.

You can also create efficiencies for reference information by scraping the current code and using edited code content itself to write the docs. When you can document thousands of configuration options this way, you can focus efforts elsewhere on helpful docs and let the robots take care of the mounds of reference descriptions.

Read the Docs describes this goal so well on their website:

> Whenever you push code to your favorite version control service, whether that is GitHub, BitBucket, or GitLab, we will automatically build your docs so your code and documentation are never out of sync.

Imagine if a restaurant chef never taught anyone how to prepare his most popular dish because they were the only person with the proper knife for the technique. Compare that scene with closely guarded secrets of proprietary docs toolchains with expensive per-seat licenses. Although we all know that certain jobs require specialized knowledge or tools, we shouldn't guard writing knowledge or tools to protect jobs.

Conversely, developer workflows shouldn't be mysterious to writers. Learning techniques while working side-by-side is a great way to demystify developer tools.

And from another perspective, developers shouldn't arrogantly think their code is all that matters, and docs don't matter because their users can "read the code." We can all think of examples of failures caused by a lack of empathy.

By treating docs as code, you open up possibilities for automation, continuous integration, deployment to web servers, and lights-out docs service for around-the-clock support.

Get value from cost-effective tools

In a publishing toolchain you need to meet several use cases: authoring (and organizing), reviewing, approving, building output, and ensuring your output meets requirements. Output requirements can vary in complexity when it comes to access control, targeted platforms, versions available, and translation needs, to name a few.

Authoring tools

The authoring tools for simple ASCII-based markup languages are often free or inexpensive. These writing tools are priced based on a per-user license, or they are completely free and open source.

Writers and developers alike appreciate using command-line editors that can be readily configured.

Side-by-side editors, which let you see the markup in one window and the rendered output in another window, are available as extensions for Integrated Development Environments (IDEs) like Visual Studio Code[10].

You can also pay about US$15 per user for side-by-side editing tools on the Mac or Windows like iA Writer[11].

Tools to review, approve, build, and manage content

In a docs-as-code system, the code version control system such as GitHub or GitLab serves as a content management system.

---

[10] https://code.visualstudio.com/Download
[11] https://ia.net/writer/

These platforms offer:

- Access control for authors
- Management of text and image files
- Review systems
- Approval workflows
- Doc builds (for example, GitHub Pages or GitLab Pages)
- Doc page and site hosting (for example, GitHub Pages or GitLab Pages)

In another scenario, such as when you include the Read the Docs or Netlify system, GitHub or GitLab provides:

- Access control for authors
- Management of text and image files
- Review systems
- Approval workflows

Read the Docs or Netlify provides:

- Doc builds
- Doc page and site hosting

These Git-backed products provide ways to collaborate using docs-as-code tools and techniques.

- GitHub[12] is a web-based user interface for collaboration and version control with Git[13]. GitHub works as a service, with a free tier for open source, public projects. GitHub pricing[14] varies based on the services you want to use, such as developer environments, packaging, automation, and collaboration features.

- GitLab[15] is a continuous integration and source control platform with a free tier and also has an open-source server that you can self-host. GitLab has many of the same features as GitHub, including free static site hosting and collaboration and continuous deployment options. An enterprise edition is also available.

---

[12] https://github.com/
[13] https://git-scm.com/
[14] https://github.com/pricing
[15] https://about.gitlab.com/

- Bitbucket[16] is part of the Atlassian[17]® family of software tracking and development products, including the Confluence® wiki tool and the JIRA® bug tracker.

Further considerations about gathering requirements for tool selection can be found in Plan for Docs as Code.

> **Query**: What about wikis?
>
> Though wikis don't gain all the automation and source control advantages of a docs-as-code framework, they do provide a simple and collaborative authoring environment with a low barrier to entry.

Examples of docs as code

A groundswell of docs as code examples has happened even in the last few years. You can find many examples to study.

In *The Product is Docs: Writing technical documentation in a product development group*, the authors mention moving their developer site to Markdown and Git in 2019 to align with and take advantage of developer tooling and build processes.

In 2020, 75% of the more than sixty Season of Docs projects used GitHub or GitLab as tools in their writing projects. The data is available as a link from the Successful 2020 Season of Docs technical writing projects'[18] previous seasons' pages.

In August 2021, the Cloud Native Computing Foundation provided scripts for analyzing growth in participation in open source projects on GitHub[19]. Their analysis[20] showed that from January 1, 2020, to December 31, 2021, Amazon Web Services (AWS) Docs was ranked number 28 with 974 contributors. Realize that this number is compared to all open source projects on GitHub, which consists of millions of projects. To have open-source docs-as-code projects on the chart at all shows how normal the practices have become.

---

[16] https://bitbucket.org/

[17] https://www.atlassian.com/

[18] https://developers.google.com/season-of-docs/docs/2020/participants

[19] https://github.com/cncf/velocity

[20] https://www.cncf.io/blog/2021/08/02/update-on-cncf-and-open-source-project-velocity-2020/

Read the Docs[21], which automates the building, versioning, and hosting of technical documentation, and serves over 55 million pages of documentation for 80,000 projects and 100,000 users. All of these docs sites use a well-known docs-as-code model and toolchain with Sphinx for docs source and GitHub, BitBucket, or GitLab for version control.

In the planning section, look into where you would push the limits of these techniques. You don't want to over-engineer the docs solutions in complex release and deliverable environments. The number of language translations and release frequencies can multiply the difficulty level for docs-as-code. You also want to be careful over promising and underdelivering in places where you need more technical support for the docs-as-code solution than you can practically get with your budget or resources. You do want to start somewhere and make incremental changes as you learn what works best for your situation.

When you want to integrate with development, collaborate with technical users and developers, and create deliverables quickly and accurately, these techniques are for you. Be smart and practical when choosing your path forward.

Background for docs as code

The desire to treat docs as code goes naturally hand-in-hand with certain environments such as developer portals, system administration guides, technical reference information, and API documentation.

Customer support site maintainers have also reported success going to a docs-as-code model. Less common examples of people using these tools are in environments where authors likely do not have a GitHub account nor a willingness to maintain a development environment.

As pre-requisite knowledge, you want to have the definition of the modern software development and modern documentation techniques and tools listed here in mind:

---

[21] https://readthedocs.org/

Software development terms

Agile development

A practice of software development and project management where scrum teams provide working prototypes and features through fast iterations that are called sprints.

APIs

Application Programming Interfaces provide ways for a developer to interact with a web service or a computer system by writing a program to instruct the system to perform an action or orchestrate a plan.

containers

A virtual image format that bundles dependencies into a single text file so that developers or machines can run an environment in a small slice of a server or local computer. Containers are often associated with Docker. Files named `Dockerfile` are a standard format to describe the base image and dependencies to install and run on the virtual instance.

Continuous Integration and Continuous Deployment

A practice of automatically building and deploying software applications or websites or other artifacts using scripts and ephemeral virtual environments for testing and staging releases, then deploying to a production environment.

DevOps

A combination of Developer and Operator, DevOps is a practice and culture of providing useful technology solutions without separating the skillsets normally reserved for the two roles. Often associated with cloud infrastructure, continuous integration and deployment, and other Agile development practices, DevOps aims to decrease release cycle times and code quality and improve application operational performance, security, and efficiency.

Git

An open-source version control system that was originally proposed by Linus Torvald to replace the version control system used to build Linux. Integrates with GitHub, GitLab, and BitBucket social coding web services.

GitHub

A company and service owned by Microsoft that offers collaborative coding features when used with Git, a version control system.

GitLab

A company providing a DevOps CICD platform that offers collaborative coding features integrated with the Git version control system.

Headless CMS

A Content Management System that makes the content visible with an API rather than with a User Interface, hence it separates the "body" or content from the "head" or presentation layer. An API lets developers choose the language and framework to build the front-end upon. The front-end can then be any device.

JAMStack

A phrase coined by the CEO of Netlify Mathias Biilmann in 2015, it stands for JavaScript, API, and Markup in a stack of applications that enables running a simple, high-performance website with the application logic on the client side rather than the server side.

linter

A tool to analyze and find problems in source code or docs based on a defined ruleset.

Netlify

A company and service set up specifically for CICD for websites that work well with the JAMStack, JavaScript, API, and Markup, a term coined by the Netlify CEO in 2015. Many technical documentation sites use Netlify for CICD due to its implementation of the static site generator build use case.

REST API

Stands for REpresentational State Transfer (REST) Application Programming Interface (API), it is an architecture style for software designs for web services accessible over HTTP on the internet. Several features make an API RESTful: resources are identified and managed with specified endpoint calls, the architecture has a separate client and server, you can make one call independently of another call (stateless), and you can send or receive JSON, XML, text, or other formats.

static site generator

A static site generator takes simple markup text files, data, and templates and transforms them into HTML, with related JavaScript, and CSS files that can be copied to a web server to create a website. You can manage templated information from data sources such as JSON or front matter metadata, and JavaScript provides search, navigation, authentication, and other interactive features. You can copy all the generated files to a web server or cloud storage system to serve as a website or docs site. Typically the files are served with a cloud storage service like Amazon Web Services (AWS) S3, an API-accessible storage REST service that can serve the URLs for the site. Common static site generators are Docusaurus, Gatsby, Hugo, Jekyll, and Sphinx.

How to learn Git and GitHub or GitLab

This book does not teach Git or GitHub. To learn Git, GitHub, GitLab, or other similar systems, rely on tutorials online and practice consistently. Choose from a few programs and match your learning style:

- YouTube video playlist, Git and GitHub for Poets[22]

- Udemy course, Git and GitHub for Writers[23]

- Official documentation, Set up Git[24]

- Book, Pro Git[25]

- Video course, How to Contribute to an Open Source Project on GitHub[26]

You can use Git at the command line or as an integrated part of your authoring environment. Common authoring tools like VS Code[27], IntelliJ IDEA[28], and Sublime Text[29] all have integration with Git as part of the interface.

---

[22] https://www.youtube.com/playlist?list=PLRqwX-V7Uu6ZF9C0YMKuns9sLDzK6zoiV

[23] https://www.udemy.com/course/git-and-github-for-writers

[24] https://docs.github.com/en/get-started/quickstart/set-up-git

[25] https://www.git-scm.com/book/en/v2

[26] https://app.egghead.io/playlists/how-to-contribute-to-an-open-source-project-on-github

[27] https://code.visualstudio.com/

[28] https://www.jetbrains.com/idea/

[29] https://www.sublimetext.com/

Git, GitHub, and GitLab terms and definitions

Familiarity with some Git, GitHub, and GitLab terms and definitions is helpful as you consider a docs-as-code transformation.

Here, the definitions tie into docs and publishing contexts.

branch
A parallel version of a repo within the repo that does not affect the primary or `main` branch (previously named `master` by default). You can work freely in a branch without affecting the *live* version. After you make changes, you can merge your branch into the `main` branch to publish your changes.

New contributors can confuse named directories, such as a cloned fork that is named after the original repo, and Git branches.

You can instruct Git to base your branch on the `main` branch in upstream, origin, or another remote. For example, this command bases a new branch on the `main` branch in the `upstream` remote:

$ git checkout upstream/main -b <branch>

clone
A copy of a repo that lives on your computer instead of on a website's server.

commit
A point-in-time snapshot of a repo. Commits let you see the differences between changes. A commit is an individual change to a file or set of files. Every time that you save a file or a set of files, Git creates a unique ID, also known as the *SHA* or *hash*, that tracks the changes. Commits usually contain a commit message, which is a brief description of what changes were made.

downstream
A label for a remote URL, where a remote represents a place where code is stored. A downstream remote indicates an opposite of an upstream, or original (origin), repo.

fork (noun)

A copy of the repo that is entirely yours in your namespace. A fork gives you a way to both contribute openly and get credit for your contributions.

fork (verb)

The act of making a forked copy of the repo.

issue

A way to submit a suggested improvement, defect, task, or feature request to a repo. In a public repo, anyone can create an issue. Each issue contains its own discussion forum. You can label an issue and assign it to a user.

merge request (GitLab)

A comparison created by Git that you send to GitLab to ask for review and potential merging to the main branch of the repository. The equivalent in GitHub is the pull request. The final action is to merge the changes to the main branch, the first action is to pull the changes to the main branch.

organization

A collection of group-owned repositories.

project (GitHub)

In GitHub, a project provides a user interface for tracking purposes on Issues with specific labels, configured by admins on the repository. Two UIs exist for Projects in GitHub in mid-2022.

project (GitLab)

In GitLab, a project contains the repository plus related issue tracking, merge requests, and CICD pipelines, in one organization structure and reference URL.

pull request

A method of submitting edits that compares your changes with the original. Teams can view the comparison to decide whether they want to accept the changes.

push

> Move your local committed changes to a remote location, such as GitHub.com[30], so that other people can access them.

remote

> A version of your project that is hosted on the Internet or on a network. The remote is usually connected to local clones so that you can sync changes.

repo

> A collection of stored code or docs.

review

> Perform a line-by-line comparison of a change and comment on improvements or suggest changes, much like a copy editor does for a newspaper article.

upstream

> The primary label for the remote URL indicates the original repo where changes are merged. The branch, or fork, where you do your work is called *downstream*.

version control

> A class of systems responsible for managing changes to computer programs, documents, large websites, or other collections of information.

Now that you have some concepts at the ready you can dig deeper into the world of docs as code.

---

[30] https://www.github.com/

Chapter 2

Plan for docs as code

> Requirements gathering

> How to choose a static site generator

> Planning for automating, testing, and site hosting

> Choosing a markup language

Plan for docs as code

Planning for any docs solution means gathering requirements from the business or product team, and your stakeholders such as end-users, product managers, business owners, contributors, or authors.

Sometimes a docs-as-code approach is already chosen for you, and when that's the case, you should still look for ways to optimize based on organizational changes, new business needs, or conversations with stakeholders. You may have some decision points along the way. The requirements gathering process helps with those decisions.

Many teams start with a simple website from a single GitHub repo. But as your content outgrows this one-to-one model, you must organize your deliverables in a meaningful way while maintaining an architecture that maximizes your ability to send review notifications and automate builds.

Requirements gathering

In an enterprise environment, you would call these needs business requirements. In an open-source setting, instead of a budget constraint, you would think of a resource constraint. That said, with sponsorships and fiscal hosting mechanisms online, the open-source world has tools to submit expenses and manage finances such as with the Open Source Collective[1].

Look out for "analysis paralysis" in the planning stage. Sometimes you need to try something, anything, and fix and improve as you go. Later you can see which static site generators or other tools fulfill which needs.

Ask some questions while you evaluate docs-as-code solutions, whether for a new system or updates to an existing one.

Budget requirements

How much should you allocate per author, per deliverable, and for hosting? Does any cost go up as scale goes up? For example, are integrations with an identity system going to cost more per user?

Are there tiered costs for any performance needs such as a global Content Delivery Network?

Per-author costs should be minimal since most Markdown systems are free or plugins to open source Integrated Development Environments (IDEs) like Visual Studio Code.

Hosting on a subdomain using something like GitHub Pages should not cost anything, and using Origin CA to generate certificates using Let's Encrypt is now free. Your company IT department would have a process to go through to set up a subdomain and probably security requirements for registering a certificate.

---

[1] https://opencollective.com/

Technical support needs for the docs-as-code system

If the system breaks, can you and the writing team fix it, or do fixes require engineering expertise?

Are you relying on an internal team or a third party to have "pager duty" if the docs site goes down?

Requirements for product versions

What domain names exist already, and do you need to fit into that structure? Do you need to build an entire website subdomain, such as docs.example.com or developer.example.com? Do you have multiple products you write docs for?

Who will integrate logins or identity management systems with the docs site? What are the requirements if you need to integrate with a knowledge base, product user interface, SEO, social media, or web analytics?

What about integrating or releasing with a software-as-a-service product, what is the cadence for releases, and what integration points indicate the release synch points?

Security requirements

Are the docs hosted on-premises or a cloud service? If hosted somewhere, what are the TLS, SSL, or certificate requirements?

What access controls are needed to protect some or all of the documentation pages?

Recent releases of Enterprise GitHub Cloud support GitHub Pages, so if Enterprise GitHub security meets requirements and limits access to employees, you could use Enterprise GitHub Pages to meet security requirements. Ensure your TLS, SSL, and certificates meet your company's InfoSec requirements, security-related companies may not allow Let's Encrypt certificates, for example.

Often you can integrate with an existing identity system rather than creating one separately for the documentation system.

Scale

Smaller docs sites or non-interactive systems may not need to worry about scale and performance, but very large docs sites need to ask, what is the expected number of users served concurrently?

If the docs are interactive, what backend infrastructure must be available to serve the correct number of users? Work with your engineering teams to help determine the right amount of infrastructure to use and also look for ways to architect a system that meets requirements at peak use times.

Brand and style needs

If you need to adjust the theme to create additional brand styling, do you have to pay periodic amounts for front-end web development, or is it a one-time investment?

Can a central team at your company help with front-end web development?

Front-end contracting companies can provide ongoing support as part of their contract, or provide documentation so the docs team can take over maintenance.

Working with an existing design team can help meet brand standards, especially when that team is already well informed on company brand and style requirements.

Migration possibilites

What is the initial amount of hours needed for migration from the current system?

Is there a clear map from one system to the other for both authors and end users?

Even an estimate for how many hours are needed to move from one system to another helps you with planning and setting expectations. Also, you have an opportunity to show how much time you can save with automation.

Platforms for docs site output

How do users access the docs, and how often from each platform?

Use any existing data you can find to estimate how users access docs today, and then prioritize creating the best experience on that platform. You can focus relentlessly and know you put the time investment to good use.

Access control requirements

Does the source repository need to be private or can it be publicly available?

Do you need to control access by a team, by role, or by user ID?

When using Enterprise GitHub licenses managed by your IT team, it's easier to manage access when someone leaves the company because all of their access to every company asset is revoked.

It's possible to use your company's existing Active Directory groups and teams to manage access. Look for integration with existing identity systems for finer-tuned access controls.

Requirements to preview the docs-as-code output

Does the authoring system provide side-by-side views of the output, rendered as it should look when built?

Can authors have a cross-platform (Windows, Mac, Linux) method to see a local preview of the docs?

Many IDEs like Visual Studio Code have an extension that provides side-by-side editing with Markdown. You can also preview the Markdown in GitHub's web view by uploading the files to GitHub and viewing them there.

Provide cross-platform support by using Docker as the development environment. That way, you manage dependencies outside of the author's local environment and in the container as much as possible.

Requirements for conditional output and variables

Do you want to build different outputs based on the targeted operating system, version of the product, or mobile or other platforms?

Do you want to use standardized Markdown for portability, or do you need to have another specific markdown language such as reStructuredText (RST) because of Python-based cross-references or extensions?

You may want to consider the programmer support you can get with the static site generator you choose. Here is a table showing the relationship between the programming language and the static site generator built in that language.

Programming Language	Static Site Generator (SSG)	Markup Languages
Python	Sphinx, MkDocs	RST, Markdown
Ruby	Jekyll	Markdown
Go	Hugo, Gatsby	Markdown
JavaScript	Docusaurus	MDX, Markdown
JavaScript	Antora	AsciiDoc
Ruby	AsciiDoctor *	AsciiDoc

* AsciiDoctor is not a Static Site Generator but a text processor that can make HTML 5, ePub, PDF, and so on.

Refer to more discussion about markup languages in "Choosing a markup language" on page 37.

How to choose a static site generator

Static site generators work great to generate web and other outputs from source files that you store in GitHub and author in lightweight markup languages like Markdown or RST. For comparisons, see Static Site Generators[2]. As of July 2022, you can choose among 460 static site generators (the list only increased by about a dozen in five years).

One modern innovation with static site generators is the headless CMS that works with static site generator output. Working with a headless CMS enables you to integrate with marketing efforts

---

[2] https://staticsitegenerators.net/

such as newsletters, social media such as Twitter cards[3], support, and other brand-related or company content or experiences that may already exist. That integration gives you more opportunity to provide a digital journey across multiple domains, apps, websites, or even devices. One example is the open-source Ghost CMS[4], developed as a Kickstarter project in 2013[5]. You need developer resources for integration with the Ghost CMS, but by providing already supported and generated content, it's possible to integrate more quickly with a headless CMS like Ghost.

Another valuable tool set is the developer portal builder TechDocs[6] which is part of Backstage[7]. Backstage is an open platform for building developer portals. It provides a wrapper for the MkDocs[8] static site generator. Spotify made it open source in 2020[9] and they continue to use it in 2022 for their developer experience portal. Their portal combines more than 5,000 documentation sites and averages 10,000 daily page views[10].

> **Query**: What if you never build the docs and serve Markdown from GitHub?
>
> GitHub provides an HTML rendering of many ASCII markup pages with GitHub Flavored Markdown as the supported syntax[11]. Many projects don't even bother to render their docs on a separate website. There is no navigation for reading source docs on GitHub unless you make the links yourself. You can get website analytics through the graphs page on your repository. Refer to this blog post, Introducing GitHub Traffic Analytics[12] for more information.

So it's possible to only write docs and never publish them to a new URL, but with minimal user experience niceties.

---

[3] https://developer.twitter.com/en/docs/twitter-for-websites/cards/overview/abouts-cards

[4] https://ghost.org/docs/

[5] https://www.kickstarter.com/projects/johnonolan/ghost-just-a-blogging-platform/description

[6] https://backstage.io/docs/features/techdocs/techdocs-overview

[7] https://backstage.io/

[8] https://www.mkdocs.org/

[9] https://backstage.io/blog/2020/09/08/announcing-tech-docs

[10] https://backstage.io/docs/features/techdocs/techdocs-overview

[11] https://docs.github.com/en/get-started/writing-on-github/getting-started-with-writing-and-formatting-on-github/about-writing-and-formatting-on-github

[12] https://github.blog/2014-01-07-introducing-github-traffic-analytics/

Approval workflows

You can set up a repository with protected branches, so that pull requests merge only after meeting certain criteria. Refer to Workflows for an in-depth discussion and decision points about workflow settings and branch protections.

Planning for automating, testing, and site hosting

In addition to using the code-hosting and review tools offered by GitHub and other code-hosting services, you can gain efficiency by automating tests and builds. Many of the integrated build services that work with GitHub or BitBucket have a free tier for public repos, with monthly subscriptions for additional needs such as builds for private repos. When you want to automate testing, you should compare pricing based on the number of jobs, repositories, collaborators, and security considerations, including whether the repo is public or private. For example, when you look into pricing for the Travis CI[13] continuous integration tool you get a month for free to start, and then for one concurrent build on a public or private repo, you pay $69 per month, for two you pay more, and so on. With Travis CI, Jenkins, or other CI platforms, you would write the script that builds the site.

Many docs-as-code aficionados have discovered the Netlify platform[14], which builds websites automatically and fits in well with the docs-as-code mindset. Since Netlify combines site previews and global deployment to a Content Delivery Network (CDN) with automated builds, it makes sense to use Netlify for building docs sites. You configure all the settings for automated builds with a `netlify.toml` file in the root of the repository. Refer to the Netlify documentation for all the details.

If you don't use a combined build and host platform, hosting the docs website might have associated costs, such as about US$10 per year for maintaining a custom domain name through a DNS

---

[13] https://www.travis-ci.com/pricing/
[14] https://www.netlify.com/

registrar. You might need to pay for web hosting itself. If you want HTTPS, you can purchase an SSL certificate for about US$100 to US$640 per year depending on the number of subdomains and possible warranty ranges. You can get free certificates through Let's Encrypt[15] or other provisioning services that provide automation and free certificates.

Using GitHub Pages or GitLab Pages for hosting has no associated cost and you can encrypt with HTTPS. Bitbucket offers a similar hosting service through Bitbucket Cloud[16] and uses HTTPS. You can use Cloudflare Pages[17] with GitHub or GitLab integration for automated site deployment with a free tier and a US$20/month Pro tier with 5 concurrent builds and 5,000 builds per month. Netlify, mentioned previously, provides a similar automated build and hosting service.

Choosing a markup language

You want to educate yourself about the background of different options since there are options beyond Markdown with the CommonMark specification[18]. The options include Markdown with the CommonMark specification, GitHub-Flavored Markdown, AsciiDoc, MDX, Markdoc, and reStructuredText (RST).

Markdown

Sometimes people think that Markdown is the first, best choice for the markup language with docs-as-code techniques. It is an excellent choice, as it is ubiquitous and the intent was to make it as readable as possible in its raw format. But did you know that the simplest original CommonMark Spec syntax does not include code blocks, footnotes, tables, specific programming language syntax highlighting, and URL auto-linking? Extensions give us those "must have" features.

We take those extensions for granted now. Having a specification allows for extensions to be built on top. One specification

---

[15] https://letsencrypt.org /

[16] https://support.atlassian.com/bitbucket-cloud/docs/publishing-a-website-on-bitbucket-cloud /

[17] https://pages.cloudflare.com/

[18] https://github.com/commonmark/commonmark-spec

published by GitHub is based on the original CommonMark spec, the GitHub-Flavored Markdown specification.

In March 2017, GitHub published a formal spec for GitHub-Flavored Markdown. The GitHub Flavored Markdown (GFM) specification provides what you read on GitHub README and other Markdown files in every GitHub repo. This publication offers a resolution to the difficulty in using Markdown for technical documentation that Eric Holscher, founder of Read the Docs discusses in a blog post from a year before the GFM spec was published[19].

From the GitHub blog post[20]: "This formal specification is based on CommonMark, an ambitious project to formally specify the Markdown syntax used by many websites on the internet in a way that reflects its real world usage. CommonMark allows people to continue using Markdown the same way they always have, while offering developers a comprehensive specification and reference implementations to interoperate and display Markdown in a consistent way between platforms." Specifically, the GitHub docs team uses the Remark processor in mid-2022 for Markdown parsing: https://www.npmjs.com/package/remark.

Now you can analyze whether a tool or parser adheres to a particular spec as part of your tool's requirements. The considerations are for portability to other platforms, or for what formats contributors may already know well.

Diagrams embedded in Markdown

For diagrams in Markdown files, there's an extension called Mermaid, at https://mermaid-js.github.io. You might want to describe a flow chart or class diagram using text embedded in the docs. Mermaid handles this use case by creating images in the output based on text chunks in the input.

An online, live editor is available at https://mermaid.live/ where you can see the diagram rendered side-by-side with the Mermaid

---

[19] https://ericholscher.com/blog/2016/mar/15/dont-use-markdown-for-technical-docs /

[20] https://github.blog/2017-03-14-a-formal-spec-for-github-markdown

graph code. This example shows a Git branch system example, called `gitGraph` in Mermaid syntax[21].

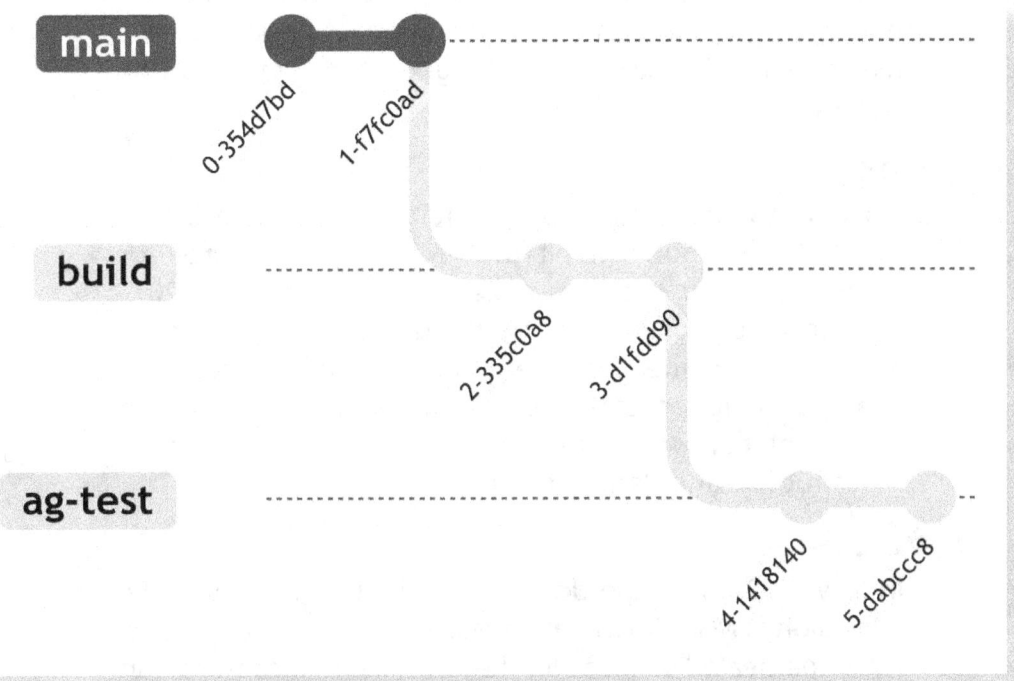

Mermaid text image gitGraph example rendered

The syntax uses a triple colon indicator or the triple back-tick and `mermaid` label like so:

```
:::mermaid
gitGraph
    commit
    commit
    branch build
    commit
    commit
    branch ag-test
    commit
    commit
:::
```

AsciiDoc

AsciiDoc provides another markup language so that writing is "as simple as writing an email" with lightweight, easy-to-memorize (and read) syntax. AsciiDoc was designed in 2002 to be a shorthand replacement for DocBook, which is a powerful XML standard used in the publishing industry. AsciiDoc can produce more outputs than Markdown "out of the box" including print, ePub, and slide layout, unlike Markdown which was never designed for print or ebook outputs.

MDX

Docusaurus is a static site generator written specifically for documentation, and it uses MDX as the source for content (and for code examples). In the December 2017 announcement blog post, the (then Facebook) developer advocate team member Joel Marcey writes, "Docusaurus is a tool designed to make it easy for teams to publish documentation websites without having to worry about the infrastructure and design details."

Markdoc

In May 2022, the Stripe docs team published their syntactic extension of Markdown, called Markdoc, at https://markdoc.io/. Stripe created Markdoc for their developer documentation and then made it open source. It's both a toolchain and a syntax. In the FAQ, the discussion about "why not MDX?" and "why not AsciiDoc?" includes the comment, "Markdoc enforces a strict separation between code and content (think: docs as data)." Markdoc is one to watch as the team provides the background on the design decisions made to bring it to life.

ReStructuredText (RST)

RST was built into Sphinx, the Python text processor, in 2002, and provides both tooling (a parser) and syntax for lightweight markup. RST is used for the Read The Docs site builds at readthedocs.org, though Sphinx can build from both RST and Markdown source files.

RST has multiple features that make it an excellent choice for technical documentation. RST supports multiple ways of building tables including CSV tables which enable you to reference table data from an external file. RST also has inclusion directives that enable the reuse of snippets or external files. You can create cross-references within a document and create variables that you can employ for those inevitable product name changes. Plus RST and Sphinx provide the ability to annotate code examples by emphasizing particular lines in code blocks. Even as a lightweight markup syntax it still packs a lot of powerful capability.

Organize source files and repositories based on team size and deliverables

How do you know when to make a single deliverable from one repo or make a large website from multiple repos? The way that you organize information influences how you can present it to readers.

To help contributors form a mental map of the docs, you might synchronize the way that you organize source files with the organization of your site. If you have a simple product, a single repo that contains both the code and docs can be the easiest. For a complex set of products or projects, you have more choices about how you organize source files and provide navigation on your site.

To decide whether to keep docs in the dev repo, use a single docs repo, or use multiple docs repos, consider your users, contributors, reviewers, content size automated tests, deployed docs, and translated docs.

Users. The way that you organize your information must not detract from the user experience. Think about how users will read each page, navigate through the site, and find what they need. To make the content useful to users, you can interview users, complete card-sort tests, and use other methods to organize your site. For example, if you decided to create books, be sure that your users actually *need* books, and that you aren't just doing it because you are used to books and they are easier to maintain.

Contributors. If developers are struggling to write docs in their repo, consider a separate repo for docs. But, if technical accuracy is the higher priority, keep the docs in the dev repo and train developers to cultivate empathy for their end-users who must understand the underlying layers of the software or service. For example, when only a handful of people know the best practices for a complex product, get those contributors together to hash out the best content they can. You'll be surprised at how lively the discussion can be when the contributors' goal is to help others understand a complex system.

Reviewers. When you create repos for docs, you must consider which collaborators are the best reviewers of the technical content. These collaborators are your main reviewers. Are these reviewers willing to set up content-change notifications in only a single repo or across multiple repos? If a cultural difference exists between docs reviews and code reviews, you must explain the docs review guidelines to your reviewers. You must also trust their judgment, especially if the continuous integration (CI) and continuous delivery (CD) processes publish the repo to the public web.

Content size. How important is it to have a comprehensive and cohesive deliverable that the entire review team knows? If all reviewers must know the entire doc base, it might be useful to put all the docs in a single repo. However, if you need deep reviews on technical content that developers know best, you can house some deliverables in a separate docs repo and use web design to deliver seamless output.

Automated tests. Can you re-use code examples in the docs that already have automated tests in the code repository? You can ensure that all the examples in the docs are working examples. Referencing files within the same repository and then re-using both the code and the tests that check that code is a win-win re-use case.

Deployed docs. If you must tightly coordinate and release the deliverables together, use a single repo with branches or tags to indicate a release point. If you need specialists to maintain some information, separate that information into a separate docs repo and assign those specialists to that repo as its review team.

Translated docs. If you must *freeze* the docs at a certain point to ensure that teams can translate without worrying about subsequent doc changes, you might split docs into a separate repo so that you can version the docs separately from the code.

Study some established docs-as-code patterns

Let's look at some patterns in use today on GitHub in both large open source projects and enterprise products that have organized their technical documentation for open contributions.

Product or project repo and a /docs folder

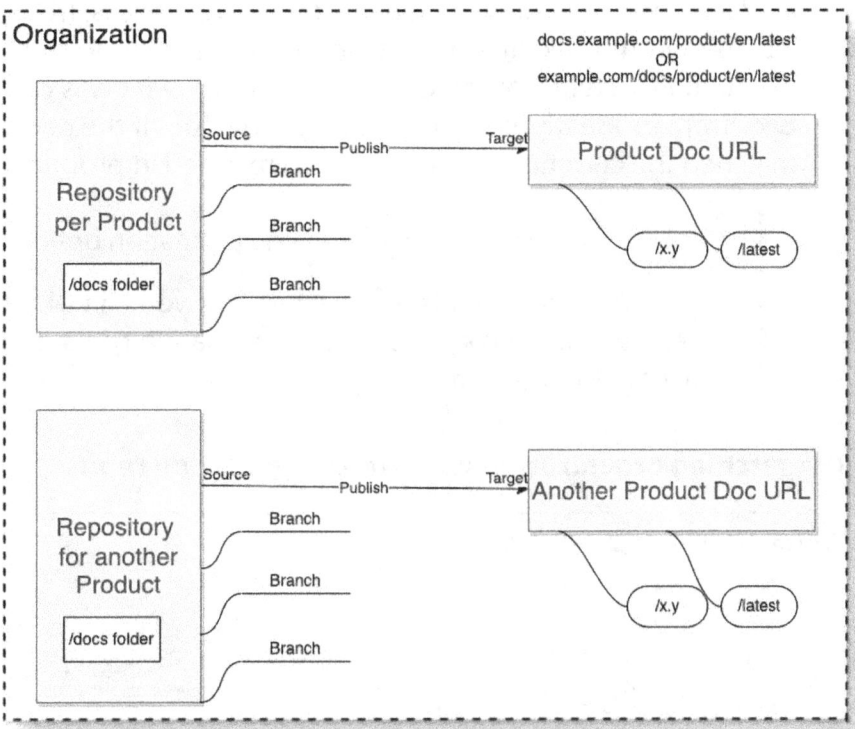

Code or product repository with a /docs folder

You could say this pattern is the original docs-as-code pattern since it's the tightest integration you can have with docs and code living in the same repository. In this case, the docs are a subfolder in the code repository and the review teams and access controls are the same for a coder and a docs contributor.

As one configuration you could set up a separate branch to publish the docs, which is a setting available in GitHub Pages. Typically this branch is called the `gh-pages` branch. You can also publish the docs with the exact branches and release tags as the code itself. This release synchronization is useful when the code and docs need to remain in synch at all times and when you are outputting to a URL that contains the version value, such as `docs.example.com/latest` and `docs.example.com/3.3`.

Any product name changes could be handled in the single repo with one pull request. In a multiple repo, multiple doc organization, you would have to clone multiple repositories if the product name change affected multiple repositories.

This pattern allows reusing any code examples in the docs that are automatically tested against the product. For example, at Rackspace we could include example responses from the REST API in the documentation, and then run tests against the production system, compare the responses to what was coming back from production, and only publish the docs if the response matched the current response that was released in production.

> **Tip**: Docs publishing workflow and automation options
>
> One advantage to this approach is that you can block merges of code changes unless the change has a doc change with the code change.

Overarching organization with website and code repos

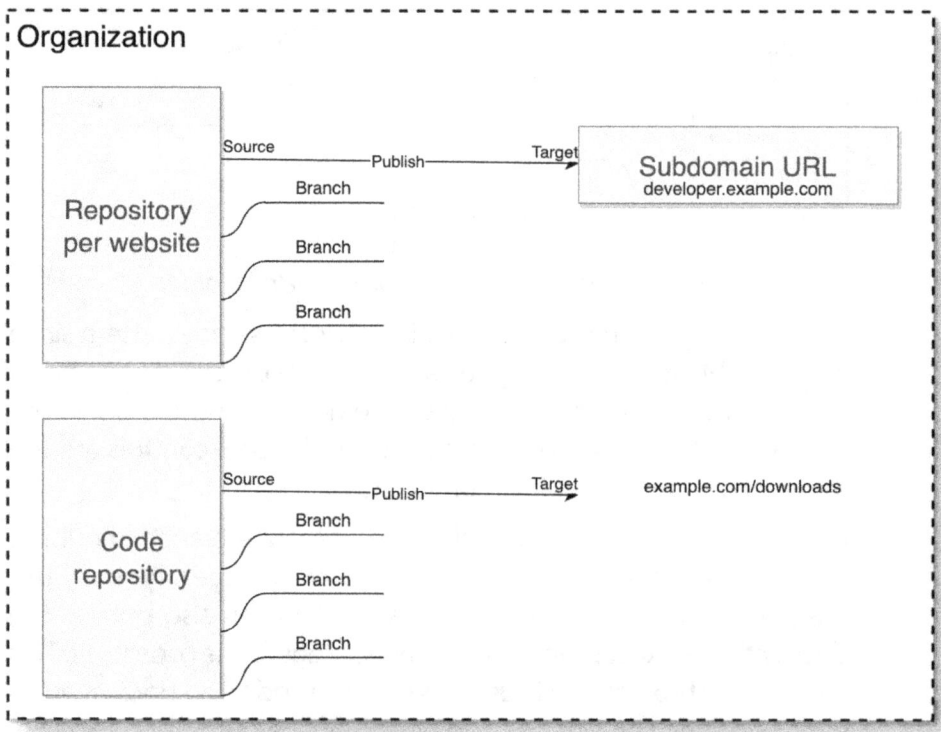

Code repos and docs repos side-by-side in one organization

In this pattern, one repo builds one deliverable, such as a docs.example.com site with all the documentation in a single repository. This pattern works well for self-contained subdomain websites like a developer portal or a single product docs site. The docs repository is independent in many ways but can follow the same workflows as other repos in the organization. The https://developer.chrome.com/ website is built from a single repository and contains docs, blogs, and articles. This works well for documenting the process simply for outside contributors to come and contribute to a single repo rather than multiple. Since it is at the same level as other code repos, possibly it will be considered similar to the code repos. Ideally, technical subject matter experts will be comfortable with the same workflows and contribution mechanisms in the docs repo.

Docs organization with doc repos

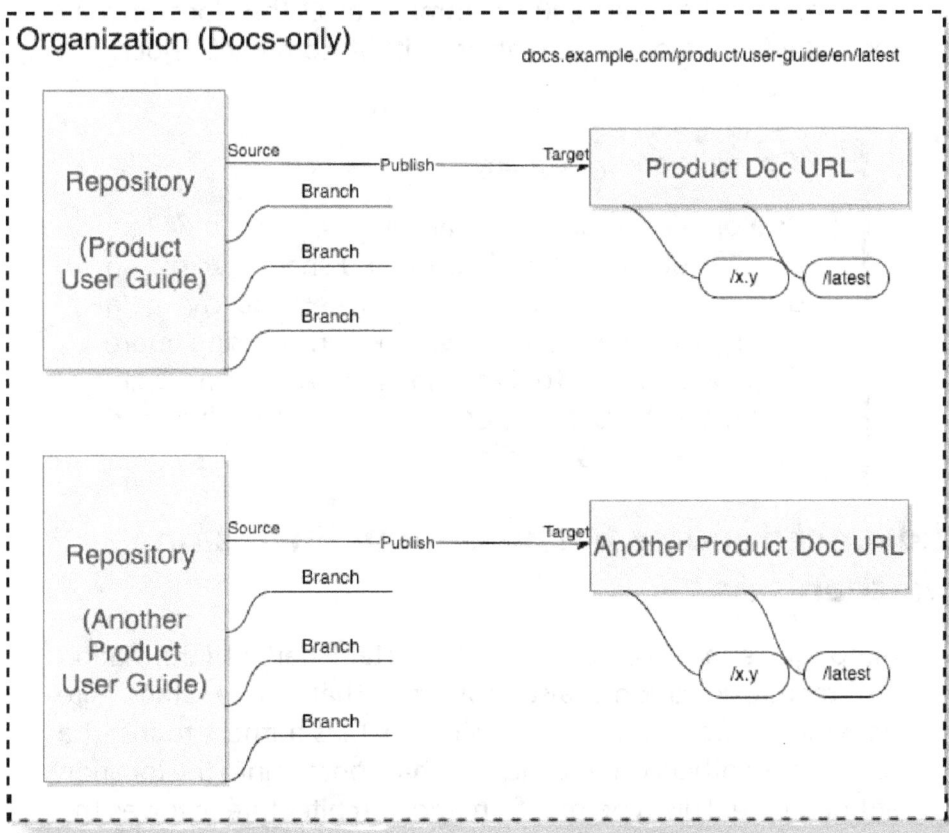

One repository per document

This organization works well when you have many products that release independently of each other, but need to have similar doc sets. You can create a template repository that people can replicate when they need to add a new document. You can publish docs faster and more frequently between major product releases by having a separate docs publishing strategy and workflow.

This approach enables product teams to set up doc review teams per repository. Teams can protect branches if needed, and give access to the repository for reviews and merges.

A disadvantage to this approach could be trying to ensure that technical subject matter experts know where to find the repos and how to contribute to these separate guides. In the case of the AWS docs at https://github.com/awsdocs, though, this does not appear to be a problem since in 2020 they had nearly 1,000 contributors and over 4,000 commits to all the docs in their repos. (Refer to reports from https://github.com/cncf/velocity for data source.)

> **Tip**: Write once, use many
>
> As a pro tip for GitHub organizations, you can write a `CONTRIBUTING.md` file once and share it across an organization. As a professional writer, you can help more than one team by writing once, using more than once. Refer to Creating a default community health file[22] for both the `CODE _ OF _ CONDUCT` and `CONTRIBUTING.md` files.

Advanced source file placement: inclusion mechanisms

Although repo placement is an important part of information architecture decisions, also consider that some static site generators offer inclusion. Inclusion enables authors to insert a reference to other content and pull that content into the location at build time. This type of information architecture involves the reuse of smaller pieces of information.

---

[22] https://docs.github.com/en/communities/setting-up-your-project-for-healthy-contributions/creating-a-default-community-health-file

For API or reference documentation, you can write excellent examples with in-line reference information. In some cases, the reference information can be maintained as doc strings in the code itself.

The Django documentation provides an example of these reference pointers and inclusion methods. Django is a Python web framework. The Django docs are maintained in a `/docs` directory on GitHub with the repo at https://github.com/django/django. Contributors use Sphinx extensions to write RST examples for Django API docs.

For example, the Django documentation contributing guide explains Django-specific markup[23].

A `ref/settings.txt` document can point to available settings with a Sphinx extension, `.. _ available-settings:`. This reference file would contain the snippet you want as the canonical text to always include.

```
.. setting:: ADMINS
ADMINS
======
Default: ``[]`` (Empty list)
A list of all the people who get code error
notifications.
```

Then a `topics/settings.txt` file could contain this when you want to reference that canonical reference snippet about the ADMINS setting:

```
You can access a :ref:`listing of all
available settings<available-settings>`.
For a list of deprecated settings see
:ref:`deprecated-settings`.
```

```
You can find both in the
:doc:`settings reference document
</ref/settings>`.
```

---

23 https://docs.djangoproject.com/en/dev/internals/contributing/writing-documentation/#django-specific-markup

Alter information and repo architecture over time

Information architecture and user experience go hand-in-hand. Treating docs as code should not limit your ability to provide well-organized docs. Take time for up-front planning and organization but also acknowledge that you might need to change repo locations, review teams, and deliverables over time.

Specialized Information: REST API docs

The docs-as-code technique provides specific wins for REST API docs.

This section discusses how the growth of REST APIs drives docs-as-code workflows that help writers work with engineers on API documentation using those standards and tools.

> **Tip**: Documenting REST APIs with the OpenAPI Specification
>
> The best way to learn about the specific standards, tools, and specifications for REST APIs is with the hands-on exercises in "Documenting APIs: A guide for technical writers and engineers" at https://idratherbewriting.com/learnapidoc/. This book does not dive deeply into how to document a REST API but that tutorial comes highly recommended.

Sometimes, an API's scale forces this collaboration. For example, when your small team of writers cannot produce docs for hundreds of API calls, you must garner the support of developers and developer experience (devex) experts with specialized knowledge.

GitHub and GitLab workflows maximize the productivity of these experts to enhance your API docs by letting multiple contributors work on the same files at the same time. Plus, when you use standard doc tools to design an API, you make a more sensible, usable API. Tools like OpenAPI, formerly Swagger, help you manage more than just the docs life cycle for a usable REST API. GitHub and version control systems are a small part of the tools' ecosystem that provides excellent API and developer docs.

To determine how many resources and which tools that you need to document an API, evaluate its complexity, the best source framework, life cycle considerations, and which user docs you need.

> API complexity

> API source frameworks

> API lifecycle

> API user docs

> API guidelines

> Studying Azure REST APIs

> Migrations for OpenStack REST APIs

> Putting it all together

API complexity

To plan appropriate contributor resources for your API docs, you must analyze the complexity of your API. An API's complexity is not determined solely by how many methods it has. A simple REST API might have ten methods but each method might have dozens of parameters.

To evaluate an API's complexity, multiply how many methods it has by how many parameters, headers, and error codes each method has. Then, match the number of assigned development and docs resources to the API's complexity.

To complete that analysis, consider these REST API components:

Component	Description
Endpoints	A REST API has endpoints to a web service that provides actions. Each endpoint could have methods to take action on the endpoint, parameters to give to the endpoint, request data or response data to or from the endpoint, and error codes that may return when trying the action with the endpoint. The number of endpoints can be a base multiplier for the size of the documentation for the API.
Methods	An API can provide HTTP GET, PUT, POST, PATCH, DELETE, and other methods. Some method implementations can be quite complex, such as a GET method for a resource collection that uses multiple query parameters for filtering and pagination.
Headers	In addition to standard HTTP request and response headers, look for any extra headers that the API defines.
Parameters	Your users must know which query, template, and body parameters are required, the default values for any optional parameters, and the constraints for parameters, such as whether a string value has a maximum number of characters. Additionally, users need to know how to use response fields from one API call to make subsequent API calls.
Error codes	Does the API use standard HTTP status codes? Are any associated error messages easy to understand? Or do you need to document additional information for each error code and message?

API source frameworks or contract specifications

After you determine your API's complexity, choose a source framework or specification for your API. An *API source framework* is also known as an *API description language* or *specification*.

It's also possible that this has been chosen for you by the development team, as the OpenAPI Specification standard has become prevalent.

The source framework that you choose helps you design the API and enables you to generate both documentation and client SDKs in multiple languages.

To specify a large REST API, ensure that you can split the long specification into separate files for ease of maintenance. Also, consider the editing environment. Try out the online Swagger Editor[24], for example.

Following are the most highly adopted frameworks:

Framework	Description
OpenAPI (formerly known as Swagger)	RESTful APIs are represented as JSON objects that conform to the JSON standards. YAML, being a superset of JSON, can be used to represent an OpenAPI Specification (OAS) file. Likely the most adopted framework, version 3.1.0 was released in 2021, and many REST APIs are documented already in version 2.
RESTful API Modeling Language (RAML)	A YAML-based language that describes RESTful APIs.
API Blueprint	A documentation-oriented web API description language based on Markdown. In addition to the regular Markdown syntax, API Blueprint conforms to the GitHub Flavored Markdown syntax.

> **Tip:** Why waddle when you can swagger?
>
> A great blog post about the origins and context of the original Wordnik Swagger specification is by Tony Tam, former Reverb CEO: Why WADL when you can swagger?[25]

---

[24] https://editor.swagger.io/
[25] https://fehguy.tumblr.com/post/10718712814/why-wadl-when-you-can-swagger

API lifecycle

When you evaluate which tools and resources you need to produce your REST API docs, you must determine your priorities for ensuring the API's success. Docs are not the *only* consideration but they are certainly an important success indicator for API use.

GitHub does not provide all the crucial life cycle systems that you need to deliver an API, though it can certainly provide data storage and retrieval, version control, and OAuth integration. Typically, an API management system provides security, parameter testing, monitoring, monetizing, visualizing, webhooks, terms of service, discovery, and the voice for the API itself. Examples include 3scale API Management Platform by Red Hat and Amazon API Gateway. SwaggerHub is an API development life cycle product with integrations. Integration starts with the OpenAPI standard itself, pointing to the importance placed on documentation for APIs.

To understand all aspects of an API life cycle, see API Transit Basics by API Evangelist Kin Lane[26].

API source frameworks fit into the larger picture of API life cycle management. REST API frameworks and companies that solve REST API management needs were snapped up in acquisitions:

- In September 2015, SmartBear Software acquired the Swagger API open source project from Reverb Technologies.
- In June 2016, Red Hat acquired 3scale, one of the early providers of API management solutions.
- In September 2016, Google acquired Apigee, an API management software provider.
- In January 2017, Oracle acquired API management startup Apiary, which supported the API Blueprint format.

All this activity points to major business drivers racing to enable APIs managed as a strategic asset. Doc platforms are a crucial piece of that modern IT delivery and support. In this business context, owning the API documentation is crucial for standards-setting and the future influence of the direction of API management.

---

[26] https://basics.apievangelist.com/

API user docs

API docs, which make the API findable and usable, are a leading indicator of an API's success. Great API docs offer a pre-sales opportunity and integration possibilities that you might not have considered.

While auto-generated reference information can provide a starting point for your users, these docs do not provide enough information to onboard users and provide ongoing support. Users must understand the API as a product that helps them accomplish a task. And the documentation is often the best representation of that product.

In addition to auto-generated reference docs, give your users a getting started guide with use cases, examples, and authorization and authentication details. You need docs that describe the architecture and why the API exists and that provide code samples and tutorials. Some great examples are the Stripe REST API docs[27] and the Parse REST API docs[28].

API guidelines

When you use GitHub and work with others on REST API docs, it's important to document your expectations for source file markup, contribution rules when documenting with code itself, completeness definitions, review guidelines, and quality indicators.

Many companies write a REST API style guide that encompasses both documentation style and REST API designs so that developers can have a consistent experience with every REST API from the company. Plus, the developers creating the APIs can better maintain, secure, and design standard APIs across engineering teams. For example, Microsoft publishes the Microsoft Azure REST API Guidelines at https://github.com/microsoft/api-guidelines/blob/vNext/azure/Guidelines.md for all their Azure development teams to use. The company-wide Microsoft API guidelines are found at https://github.com/microsoft/api-guidelines/tree/vNext.

---

[27] https://stripe.com/docs/api
[28] https://docs.parseplatform.org/

These guidelines are for:

- Consistent naming and casing in the endpoint URL and headers.
- Object collection, filtering, and pagination patterns.
- Explainable and understandable versioning for API changes and updates.
- Many more complex design decisions when making an API.

Arnaud Lauret, the API Handyman, started a website called the API Stylebook that has a collection of design guidelines and design topics at http://apistylebook.com/. It's interesting to see how many companies have published their style guidelines over the years and whether a company has dedicated resources to maintaining the guidelines. For example, Adidas, the shoe and sportswear company, has published guidelines for the API design and development at Adidas.

Checking a REST API Specification against guidelines

You can automatically check whether an OpenAPI Specification file meets set guidelines by creating a ruleset and using a linter. Developers use a linter to check if their code meets the style requirements for consistency and code rules. Similar to how a clothes dryer lint trap can capture the extra fibers from fabric while tumbling the cloth, a code linter can trap all the extra "fluff" that doesn't help with compatibility, portability, or readability of code.

As an API doc writer, you can use that same tool, a linter, to check to see if your OpenAPI spec file meets your company's API style guidelines and consistency requirements.

> **Tip**: Prose linters
>
> You can also use prose linters such as Vale[29] to check against your "regular" style guide with rulesets created for your company or team style and consistency guidance. Read more about using linters for that purpose in the testing and automation section under Optimize Docs-As-Code Workflows. You can find many example style guides in the Write the Docs > Documentation Guide > Style Guide section[30].

Studying Azure REST APIs

Microsoft has written excellent external contributor guides - or repurposed internal guidance for external reading. Either way, it's worthwhile to read Authoring good descriptions in Swagger 2.0[31] in full. Here are some takeaways.

What you write in the OpenAPI (Swagger) specification file gets output in the developer documentation word-for-word, so it's worthwhile to use content tools or content specialists to help with writing descriptions.

Visual Studio Code has helper extensions for editing and previewing Swagger2.0 and OpenAPI specifically.

Look for the free Grammarly spelling and grammar checker and use its suggestions.

It's okay to repeat yourself when writing descriptions because people will scan the document for consistent descriptions written in plain language. So writing a description, "Gets the latest alerts." is fine for `GET <path>/alerts/latest` because people read the sentence faster than they can parse the code line.

Use the style guides appropriate for your company and products. In the case of the Azure REST API specifications, they use the Microsoft Writing Style Guide[32] and the Microsoft Azure REST API Guidelines[33].

---

[29] https://github.com/errata-ai/vale

[30] https://www.writethedocs.org/guide/writing/style-guides/

[31] https://github.com/Azure/azure-rest-api-specs/blob/master/documentation/swagger-authoring-descriptions.md

[32] https://docs.microsoft.com/en-us/style-guide/welcome/

[33] https://github.com/microsoft/api-guidelines/blob/vNext/azure/Guidelines.md

The Azure REST API documentation set consists of thousands of REST API endpoints. Each endpoint has these sections generated automatically from the specification:

- URI Parameters
- Request Body
- Responses
- Security
- Examples
- Definitions

Where the verb and endpoint example is

```
PUT https://management.azure.com/
subscriptions/b67...da3/
resourcegroups/1ac...12G/providers/
Microsoft.Insights/alertrules/ch..in?api-
version=2016-03-01,
```

the example response in JSON (for that version value) is:

```
{
   "location": "West US",
   "tags": {},
   "properties": {
     "name": "ch...in",
     "description": "Pura Vida",
     "isEnabled": true,
     "condition": {
       "odata.type": "Microsoft.
Azure.Management.Insights.Models.
ThresholdRuleCondition",
         "dataSource": {
           "odata.type": "Microsoft.
Azure.Management.Insights.Models.
RuleMetricDataSource",
           "resourceUri": "/subscriptions/
b67f...da3/resourceGroups/Rac...
RG/providers/Microsoft.Web/sites/
leoalerttest",
           "metricName": "Requests"
         },
         "operator": "GreaterThan",
         "threshold": 3,
```

```
      "windowSize": "PT5M",
      "timeAggregation": "Total"
    },
    "actions": []
  }
}
```

This special version value response insertion in the documentation happens because of the "x-ms-examples" extension entry in the Swagger file[34] and a $ref cross reference pointer, or reference to a definition in another JSON file. This $ref reference syntax[35] enables you to maintain specification files in multiple files and point to them rather than having one long JSON file for the specification.

The Azure REST API documentation folder[36] provides many opportunities to learn.

Migrations for OpenStack REST APIs

The Web Application Description Language (WADL) is an XML-based standard submitted to the World Wide Web Consortium (W3C) in 2009. In 2010, OpenStack opted to use the WADL format to document the OpenStack-defined Object Storage and Compute REST services.

XML tools specialists at Rackspace ensured that OpenStack contributors could validate developers' requests for the cloud service and keep the docs true to the implementation. The format and tooling provided a win-win situation.

In 2014, the WADL specification had not been adopted as a standard, the Rackspace doc-specific tools team was disbanded, and the ability of a single toolset and WADL-based format to filter requests and document the services proved unwieldy. Plus, the community was unable to make required doc updates by using the unnecessarily complex XML- and WADL-based toolset.

The OpenStack team recognized that it needed a new API reference doc solution. For a full year, Anne worked with others

---

[34] https://github.com/Azure/azure-rest-api-specs/blob/main/documentation/x-ms-examples.md

[35] https://swagger.io/docs/specification/using-ref/

[36] https://github.com/Azure/azure-rest-api-specs/tree/main/documentation

on ways to transform the API documentation. These were the main problems:

- Historically, API reference information has been difficult for community and company resources to maintain due to the specialized nature of the toolset and the REST API.

- API reference information must live where API writers, API developers, contributor developers, and the API working group members can review it together.

To solve those problems, developers tried to migrate the dozen or so WADL files to OpenAPI (Swagger). The partial migration left the references incomplete. People also tried to use OpenAPI to manually write the rest of the specifications. For some APIs, this tactic could work, but not for all OpenStack APIs.

After the team looked at OpenAPI as a way to document a large API like Compute (over 900 methods), it decided that OpenAPI could not be retrofitted for an already-existing API. Instead, the team opted to use reStructuredText (RST), Yet Another Markup Language (YAML), the Sphinx build framework, extensions (os-api-ref[37]), and web development in a Sphinx template (openstackdocstheme[38]) to get decent display and interaction with the reference parts. And, by using these standard formats that OpenStack developers already used heavily, the team made the process smoother for developers to contribute to the REST API docs. As an example, look at the RST source[39] for a single REST API method from 2017:

```
List Servers
============
.. rest_method:: GET /servers

Lists IDs, names, and links for all
servers.
```

The `.. rest _ method::` points to a Sphinx extension that creates the HTML and interaction for the web page.

---

[37] https://opendev.org/openstack/os-api-ref
[38] https://opendev.org/openstack/openstackdocstheme
[39] https://opendev.org/openstack/nova/src/branch/master/api-ref/source/servers.inc

Here is an example snippet of the RST that points to YAML to document the request parameters:

```
Request
-------
.. rest_parameters:: parameters.yaml

   - limit: limit
   - marker: marker
   - sort_key: sort_key_server
   - sort_dir: sort_dir_server
```

The parameters.yaml[40] file describes the key_name_query_server[41]. The output is in a nice, readable HTML table, and because `required` is marked `false`, the output indicates that it is an optional parameter.

```
key_name_query_server:
  description: |
    Filter the server list result by
keypair name.
    This parameter is restricted to
administrators until microversion 2.83.
    If non-admin users specify this
parameter on a microversion less than 2.83,
    it will be ignored.
  in: query
  required: false
  type: string
```

In 2017, as Anne looked back at the two-year REST API migration effort, she had confidence in the final solution for this difficult area of documentation.

By using a format and toolset that the community can maintain and that can produce better web experiences, the OpenStack team can offer the most accurate docs across more REST API services.

---

[40] https://opendev.org/openstack/nova/src/branch/master/api-ref/source/parameters.yaml

[41] https://opendev.org/openstack/nova/src/branch/master/api-ref/source/parameters.yaml

Putting it all together

Docs for specialized use cases, like REST API docs, require careful analysis and planning. Only moving to a docs-as-code model does not guarantee the success of your API docs. The right approach is to design your API with standards and guidelines including docs, plan your docs, and then continue to check those standards and guidelines as you iterate and improve the API docs. The OpenAPI Specification has been widely adopted by teams who want to design and document REST APIs accurately and efficiently.

Compare docs-as-code solutions to wikis

Although wikis don't meet all the criteria of a docs-as-code framework, they do provide a simple and collaborative authoring environment with a low barrier to entry. That said, GitHub and other collaborative systems built on `git` provide project wikis. Often projects use those wikis in a docs-as-code way, where files are checked in to publish them to the wiki.

As an interesting middle ground, the docToolchain project at https://doctoolchain.org/ has created a pipeline for publishing generated HTML pages to Confluence.

Using a project wiki within a Git repo may make sense for your project, whether hosted on GitHub, BitBucket, or GitLab. Nick Volynkin, a technical writer at Plesk, makes these excellent points while advocating for project wiki use in a docs-as-code environment.

- Project wikis offer an easy onboard for contributors, and are a viable choice for relatively small projects. However, as a project evolves, you likely need a full-fledged documentation website.

- Wikis can be useful for rare use cases or edge test cases, and provide troubleshooting tips for those cases. In the project wiki, you can place the topics that are not relevant to most readers or have no place in the main documentation.

Wikis can also host working drafts, team jokes, project folklore, or simple feedback pages. These pages wouldn't have to meet the standards and style guide of the main documentation, but can certainly add to the project's sense of identity.

> Usability

> Security

> Statistics

> Automated builds

> Review workflow

> Web design

> Search

> Reuse

> Build Confluence wiki pages from AsciiDoc

> Finalizing the comparison

So, why not use a wiki instead of GitHub as an authoring and content management system? This section compares the systems in some key areas.

Usability

Wiki	Many developers never leave their Terminal window and do not want to open a browser window to log in and edit a wiki page that they must find in the first place.
GitHub	You can avoid the context switch from coding to opening a wiki page by placing the docs directly in the code or the same GitHub repo as the code.

Security

Wiki	Sometimes wikis are seen as ideal internal-only documentation sites because the wiki is accessible behind a firewall, for example.
GitHub	GitHub Pages from GitHub.com are publicly available when published from a public repo. When published from a private repo, you can control access if you are using Enterprise GitHub Cloud. As an alternative to having GitHub control access, you can implement an authentication workflow on a site that you upload to GitHub Pages, but you must have the web development resources to maintain the login requirement. You could use an instance of GitHub Enterprise configured to require a VPN connection to view pages.

Statistics

Wiki	Some wiki engines give you statistics that help you determine who is a topic expert. Others may only provide statistics to users with certain permissions.
GitHub	GitHub enables you to see which contributors are working on a particular part of the software project, which can help with documentation needs.

Automated builds

Wiki	Many wikis only provide a simple Save button per page. With additional plugins for Confluence, for example, you can automate parts of the publishing process and put a workflow for reviews in place.
GitHub	Treating docs as code lets you automate more tasks than simply rendering and publishing HTML. Automated builds for review purposes are a big advantage of treating docs as code.

Review workflow

Wiki	Many wikis only provide a simple Save button per page. With additional plugins for Confluence, for example, you can automate parts of the publishing process and put a workflow for reviews in place.
GitHub	Treating docs as code lets you automate more tasks than simply rendering and publishing HTML. Automated builds for review purposes are a big advantage of treating docs as code.

Web design

Wiki	If your web development resources know a particular wiki framework well, they might be able to create a nicer end-user experience for a wiki than in a highly flexible web framework like Sphinx or Jekyll.
GitHub	When publishing an entire site using static site generators, you get more flexibility in choosing web designs and navigation than with most heavily opinionated wiki frameworks.

Search

Wiki	Wikis have search capabilities and scoped searches that make sense for projects and teams that cannot be easily replicated in GitHub repositories.
GitHub	With GitHub, you can use the search form and create a query but it is for repositories and code, not for content like documentation.

Reuse

Wiki	Wikis are page-oriented, and to do releases, you might need to publish a page twice. Over time in a wiki, contributors create new pages instead of looking for and editing existing pages. As a result, trusted content becomes harder to find. Sprawl can be a problem with both solutions, but the organized nature of having documentation near code seems to keep the two in lockstep when they share systems such as version control and review workflows.
GitHub	With GitHub, you can back port a change from one release to another by using the same pull request workflow.

Build Confluence wiki pages from AsciiDoc

The docToolchain team[42] has put together a command-line wrapper tool that can publish from AsciiDoc source to the Confluence wiki. This method provides a wiki front-end, Confluence, while offering the docs-as-code workflow and AsciiDoc markup language that tech writers appreciate.

The premise is that you install `docToolchain` as a wrapper command-line script that you can put into your automation workflow to publish when your documentation changes. Refer to the https://doctoolchain.org/ website to learn about the latest release and how to configure and automate publishing to Confluence with a docs-as-code approach.

Finalizing the comparison

To summarize the comparison between wikis and docs as code techniques, each has comparable benefits in collaboration and simple markup languages. But for bulk editing, version control, and automation efforts, docs as code has an advantage over wikis. If you want the ease of a front-end and log-in system already available in Confluence, you could use the `docToolchain` build system for automation and version control while building wiki output in Confluence.

---

[42] https://doctoolchain.org/

Chapter 3

Optimize workflows with docs as code systems

> Comparing workflows and situations

> Coach authors and give style guidance

> Automate builds so you can focus on writing

> Test the docs: linting, inclusive language, and DocOps

> Review your docs as code

> Versions and releases: publish docs as code

Git has many workflows to match the way that your team wants to work, whether you use GitHub, GitLab, Bitbucket, or another Git-based system. If your team either wants to start treating or is already treating docs as code, the workflow choice might already be made.

The following basic workflows in GitHub, for example, help you collaborate with others in your repo. They are ordered from the simplest to the most complex.

Centralized. Work in a central branch, typically named `main`. Always push changes to that branch when working with others.

Feature branch. All feature development takes place in a dedicated branch instead of the `main` branch. This encapsulation makes it easy for multiple developers to work on a particular feature without disturbing the main code base. It also means that the `main`branch never contains broken code, which is an advantage for continuous integration environments.

GitHub flow. Work in feature branches, but have agreed-upon guidelines for branch interaction. One such documented workflow is the GitHub flow written up in the GitHub docs[1].

Forking. Work independently in a fork of the repo. Merge as often as you like to your fork. Request merges through pull requests to the centralized repo (`main`) to integrate features.

> **Note**: The GitHub flow workflow uses feature branches, but you can also combine a centralized or forking workflow with feature branches. You do your work locally in a dedicated branch instead of working in the `main` branch. Then, when you create a pull request, you can request to merge your feature branch to the upstream `main` or another branch, depending on your workflow. A trunk-based workflow has you merge small, possibly frequent changes to a "trunk of the tree" or a default branch, usually named `main`.

For more information, see the Comparing Workflows[2] tutorial from Atlassian.

If the docs are in the code repo and that repo has a workflow, use the existing code workflow. If your docs are in a separate repo with a different team of reviewers, work with your team to choose a workflow.

Realize that the further the docs are from the code, the more difficult it can be to keep in sync. Think about creating a culture where code changes don't merge without corresponding doc changes. For example, use the settings in a GitHub repo that enable doc builds from a `/docs` directory on the `main` branch, which would then allow teams to review and require docs changes with any code changes before merging.

---

[1] https://docs.github.com/en/get-started/quickstart/github-flow
[2] https://www.atlassian.com/git/tutorials/comparing-workflows

Your team's history and culture can partially dictate workflow choices. For example, does your team use a locking version control system, where the system locks entire files for one person to work on at a time? Or, is the team well-tuned to the optimistic-merge view and using Git natively, where merges are considered fairly easy to resolve? When you write docs as code with a new team and new repo, you can start with a simple workflow and move to a more complex one.

Does the team have a great review culture, where the members help each other and even patch each other's patches? Or, is the team more proprietary, where each member owns a set of content that no one else touches? With a team that prefers ownership, you can use *feature branches*.

Technology constraints also dictate the workflow. Is there a technical reason for release code to be protected from the public for a while? You might need to have people working independently for a while, so *forking* makes sense. Is the centralized code base complex? Is it in a single repo or does it split across many repos? With a central branch, you might want to publish early and often, so protective measures aren't necessary.

The hardest part about choosing or switching workflows is knowing when you can, or can't, merge to `main`. It's easy to develop *muscle memory* and type `git merge origin main` even when you do not want to do that!

Comparing workflows and situations

Use the following decision table along with the information that follows the table to decide which workflow might work best for your team.

Situation	Centralized	Feature branch	GitHub flow	Forking
Are the docs in their own repo?				
Docs in separate repo	✓	✓	✓	✓
Docs in code repo	Use dev workflow			
How will you publish?				
CI publishing	✓	✓	✓	✓
GitHub Pages publishing with gh-pages	✓			
Will contributors preview docs locally or on a server?				
Preview docs locally	✓			✓
Preview docs on server	✓			✓
How will you collaborate with others?				
Low collaboration	✓			
High collaboration		✓	✓	✓
How many pull requests and reviews do you expect?				
Small number of PRs	✓			
Large number of PRs		✓	✓	✓
How often do you release your docs?				
Docs release with code	Use dev workflow			
Docs release continuously	✓	✓	✓	✓

Are the docs in their own repo?

If docs are in their own repo, you can version and release them separately from the code. You can choose a centralized, feature branch, or forking workflow.

If your docs team can choose its workflow, consider keeping it simple and using a centralized workflow. With docs in their own repo, you can set up access control so that the review system allows only the lead developers or lead writers to publish, for example.

Also, if you are using GitHub, consider using the `main` branch only for both editing and publishing. This tactic works well if you are constantly publishing and want the `main` branch to consistently reflect the current state of the docs. GitHub has three settings for publishing docs, so read Configuring a publishing source GitHub Pages[3] for details as you research workflows.

When docs are integral to the code, the docs workflow must match the development workflow because the collaborators are using the workflow that is best for that repo. You can still advocate for stable branches in the code repo for docs only, and with branch publishing, you can publish only the docs directory.

Beyond just the workflow, you must also consider the information architecture choices that individual repos give you. If the repo makes a single website, your architecture depends on the way the content is rendered. If multiple repos create content for a single website, you must have a web design that enables readers to view output from multiple repos.

How will you publish?

Without any sort of outside continuous integration (CI) tooling, you can use the `gh-pages` branch that GitHub provides for actual publishing. Using the `gh-pages` branch means that your workflow needs to include merging to `gh-pages` at specific times for publishing purposes.

In repos that represent online docs, you can make the `main` and `gh-pages` branches identical with a simple configuration setting. Then, you treat `gh-pages` as if it were the `main` branch, and you continuously publish with a simple centralized workflow. You can change which branch is used for publishing with the Settings tab for the repo.

You can publish in this way only if you are using Jekyll plug-ins that support `gh-pages`. If you have plug-ins that `gh-pages` can't publish, you must build on `main` and copy the files from

---

[3] https://docs.github.com/en/pages/getting-started-with-github-pages/configuring-a-publishing-source-for-your-github-pages-site

_ site as the root of the gh-pages branch. That said, you are required to build from the main branch when building a user or organization site. For details, see GitHub Pages docs[4].

Will contributors preview docs locally or on a server?

With either a centralized or forking workflow, you should determine how contributors can view rendered docs while working on patches. Setting up a staging server could be time-consuming and require development resources. Can contributors easily stage changes, such as by using GitHub Pages? For example, could contributors set up their forks to publish to http://username.github.io/repo-name to preview their changes, without affecting SEO or the domain name for the main site? If your docs repo is a private repo, are you willing to let the staged pull requests be found on the Internet prior to actual publishing? If not, then you likely don't want to use the forked gh-pages method for a staging server for your docs site.

Instead, provide contributors with instructions for building locally so that they can preview their changes before submitting a pull request. For example, the Apache jclouds docs build each pull request to a Rackspace Cloud Files staging location for a preview of the patch changes. A message from the build job tells the contributor how to preview the doc change[5].

With a centralized workflow, contributors need to see a preview before pushing to main. With a forking workflow, ensure that the preview meets the privacy needs of the source docs.

How will you collaborate with others?

Even with a large team, you can give contributors access to the publishing branch as long as your contribution guidelines are very clear. If you don't expect many reviews, you can pick a simpler workflow. If you expect many contributors to add features to the docs, you might choose the feature branch workflow.

---

[4]   https://pages.github.com/
[5]   https://cwiki.apache.org/confluence/display/JCLOUDS/
      How+to+Contribute+Documentation

How many pull requests and reviews do you expect?

Whether you expect to review 20 pull requests daily (that's a lot!) or one each week or month, you might want to track what is merged. If you have a small team with only a few reviews a week, it's fairly easy to remember which topics have changed or how the overall site itself has changed in a given time period. In this scenario, your workflow choice can be a simpler one.

If you have a lot of contributors and want to keep the published site fairly up-to-date, you likely want a staging environment and a production environment with branches for each. In this scenario, the feature branch workflow or a trunk-based workflow would work well.

How often do you release your docs?

If your docs must be released only at the same moment as the product is available, you likely want to merge to a separate branch for the duration of the work leading up to the release. With this consideration, a forking workflow works best for release docs.

If you must publish the site as soon as a change lands, you can use a simple workflow and even delete the default branch and use just the `gh-pages` branch.

You can also take a hybrid approach as your docs set grows. For example, let's say a product has release deliverables written for a single release every six months and continuously publishes deliverables written to span multiple versions. Release-specific books, such as the installation guides and configuration guide, reside on two currently supported branches (the current release and two previous releases). Ongoing writing work continues on the default branch. The continuously released books, such as a user guide and administration guide, are built only from the `main` or default branch.

Do you want to limit doc publishing with protected branches?

You may want to protect branches and there are several configuration settings in increasing amounts of protection available on GitHub. Branch protection helps when you want to publish only when the docs are ready for release with a product release, for example, or when you want the docs team to have the final review before publishing to production. You can configure the protection so that either a single approver or core team approves the merge. You then need to have an automatic script that publishes to production after approval. Refer to the reviews section for more details on protected branches.

Team etiquette an ideal pull request or merge request to help with reviews

You want to practice some judgement when sizing up how many changes and how many files to include in those changes.

Keep the number of files changed to less than ten for certain. Ideally, change only two to four related files.

The ideal number of lines of changes vary depending on how far apart the lines are, and if the new section is completely new. A new section can be easier to review than edits to an existing section.

You may want to put organizational changes, such as table of contents additions, movements, or deletions, in a single pull request, without any edits to the content itself. You make it easier on a reviewer when you only need a review on the organization.

Fix one doc bug per pull request so that the reviewer can focus on whether that doc bug is fixed with the change.

Finalizing your choice in workflow

The workflow that you choose depends on the history of the team and repos where they have been working. At first, a simple workflow, such as the centralized workflow, can work well. As the number of contributors, reviewers, releases, and publishing locations grows, you can add complex feature branches, use a `CODEOWNERS` file and protected branches, or fork-and-pull workflows. The amount of automation you add to the workflow can add trust to reviews as well as make a reviewer's job less time-consuming. Some rules require more work from the author, some rules require more work from reviewers.

Coach authors and give style guidance

We are here to write, after all. We author content to convey concepts, teach others, document how to do an important task, or describe required reference information.

Choose an editor

To author the source files that you store in GitHub or another code system, use a lightweight markup language such as AsciiDoc[6], Markdown[7], or reStructuredText[8] (RST). You may have reasons to choose MDX[9] or Markdoc[10] formats, which you can read more about in "Plan for docs as code."

To enable many collaborators, choose a platform-independent, text-based editor. You can invoke many text editors from the command line in a Terminal window or in a web browser.

Authors who are accustomed to an integrated development environment (IDE) find that they can configure plug-ins for docs markup languages. Visual Studio Code supports Markdown previewing by default[11] (press ⌘K V) and has extensions for more related functionality. Others who are constantly in their Terminal window and don't want to context-switch to a separate editor find `emacs` or `vi` to their liking and work on docs without losing context or switching tools.

---

[6] https://asciidoc.org/
[7] https://commonmark.org/help/
[8] https://docutils.sourceforge.io/rst.html
[9] https://mdxjs.com/docs/what-is-mdx/
[10] https://markdoc.io/docs/overview
[11] https://code.visualstudio.com/Docs/languages/markdown

You can use side-by-side editors to author simple ASCII markup and get an approximation of the output. Typically, you must be a bit skeptical of the accuracy of that output, especially for Markdown because it has a few interpretations. Still, many writers prefer the side-by-side editing that tools like Macdown[12] for Markdown and the Online reStructuredText editor[13] for RST offer. The online RST editor does not show all extended semantic markup exactly how your CSS output does, but it's a good working environment for many people. This feature is especially helpful when you are editing tables in simple markup, which is a challenge.

Build locally first

When you author content, you want to see the output as your end users will see it. As mentioned in the preceding section, some side-by-side editors let you view the output as it appears in a web browser, but that output is not always reliable.

Ideally you should have a staging environment for the docs. When you first start treating docs as code, though, you might not have the tools in place for a staging environment. If you don't have a staging environment, you must be able to build docs locally. With Go-, Ruby-, and Python-based environments, you need a development environment.

On macOS, use Homebrew[14] to manage packages and depenencies. Homebrew also has nice features like the `brew doctor` command to help troubleshoot problems with permissions or symbolic links. Using `brew` ensures that your system Python, for example, remains untouched, while you can run different versions of Python.

> **Tip:** Separate doc tools repositories
>
> Sometimes you see separate doc tools repositories, such as one for Microsoft docs that provides a Visual Studio Code extension. Those tools can be useful to authors and you can invite the community to help with tools as well.

---

[12] https://macdown.uranusjr.com/
[13] https://snippets.documatt.com/
[14] https://brew.sh/

Example source file and outputs

These examples show an RST source file and two types of output. Python developers use RST to document their code inline. They use RST-based rendering engines, such as Sphinx, to render web content.

RST source file

This example shows an RST source file. When reviewing markup, you can see what has changed. Also, the blank space in RST is as meaningful as the text itself.

```
.. _install-client:

Install and use the murano client
~~~~~~~~~~~~~~~~~~~~~~~~~~~~~~~~~~~

The Application Catalog project provides a
command-line client,python-muranoclient, which
enables you to access the project API.
For prerequisites, see `Install the prerequisite
software <http://docs.openstack.org/cli-reference/
common/cli_install_openstack_command_line_clients.
html#install-the-prerequisite-software>`_.

To install the latest murano CLI client, run the
following command in your terminal:

.. code-block:: console

    $ pip install python-muranoclient

Discover the client version number
----------------------------------

To discover the version number for the python-
muranoclient, run the following command:

.. code-block:: console

    $ murano --version

To check the latest version, see `Client library
for Murano API <https://git.openstack.org/cgit/
openstack/python-muranoclient>`_.
```

RST source file example

GitHub-rendered output

You can view an RST source file as rendered content on GitHub, as shown in this example. Markdown is also prevalent when working on GitHub, and might render better than RST in some cases. Still, note the copy to clipboard icons when mousing over the code highlighted text.

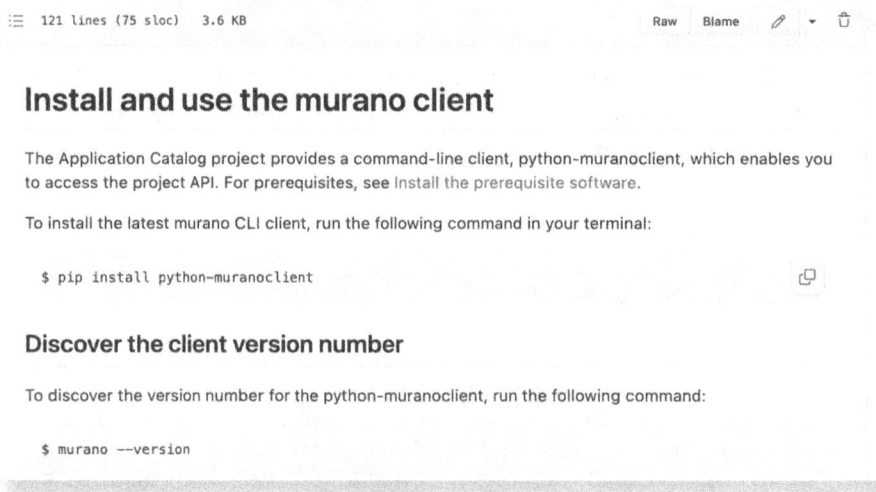

Rendered on GitHub

Website-rendered output

In addition, you can view the rendered content on a website, such as the OpenStack site, as shown in this example. This website rendering is done with a Sphinx theme.

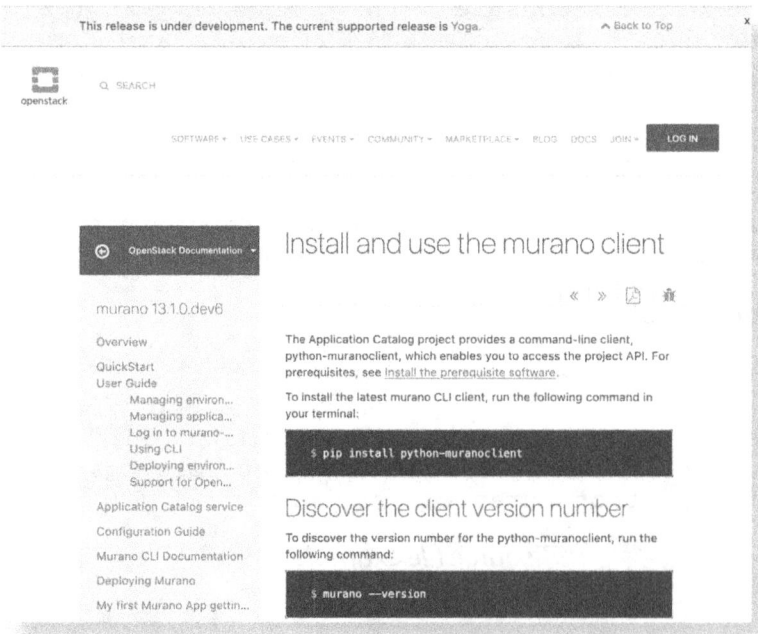

Output on web site

Writing a docs contributor guide

You want to guide both new and ongoing contributors to the docs when using docs as code. A common practice is to use a CONTRIBUTING.md file in the root of the repository, and GitHub has even provided some integration with the user interface that points to that file.

For more in-depth docs contributor guides, you likely want to publish beyond the repository. Here are some sections you want to include and questions you want to answer. Beyond the list, read some excellent specific docs contributor guide examples to get started.

- Process-based prerequisites to contribute, such as a Contributor License Agreement. A Contributor License Agreement spells out the terms under which someone has contributed the content to a project. Your legal department would help with setting up any legal or copyright needs for your docs as code projects.

- Process information, such as whether you must only work on an Issue or get an idea approved before starting work. You also must describe the desired git workflow, branch naming conventions or labels, and pull request guidance.

- Technical prerequisites to contribute, such as the authoring and development environment you need to set up first before you can begin work.

- How to get the source files and document templates.

- How to build the documentation site or output.

- Formatting information for the source files.

- Style guidance for the writing, information architecture for the organization, and topic or type guidance such as where do tutorials go? What about images or image formats, and should source files for images go in the repository?

- Set publishing and review expectations, such as timing for publication and release cadence of the documentation. Should a contributor expect to see their change immediately when their pull request is merged? Should they expect a couple of days turnaround on reviews, or is a couple of weeks more the norm?

- How much testing happens for the docs? Are there automated build checks and are the checks required or optional before merging?

- Translation information related to the docs, if it's relevant.
- How to get help if you have questions about contributing?
- It's also helpful to let people know why contributing to the documentation benefits lots of people.

Example documentation-specific contributor guides

These example contributor guides provide lots of great ideas and inspiration for your guidance.

- AWS Documentation Guidelines for Contributing - EC2 User Guide[15] Published as a CONTRIBUTING.md file[16]
- Django Writing Documentation[17] source[18]
- GitHub docs contributing guide[19] Published as a CONTRIBUTING.md file[20]
- GitLab Documentation guidelines[21]
- Google Chrome Developers Documentation Handbook[22] source[23]
- Kubernetes docs and website contributor guide - Contribute to K8s docs[24]
- Microsoft Docs contributor guide[25] source[26]
- New Relic Writing docs[27] source[28]

---

[15] https://github.com/awsdocs/amazon-ec2-user-guide/blob/master/CONTRIBUTING.md
[16] https://github.com/awsdocs/amazon-ec2-user-guide/blob/master/CONTRIBUTING.md
[17] https://docs.djangoproject.com/en/dev/internals/contributing/writing-documentation
[18] https://github.com/django/django/blob/main/docs/internals/contributing/writing-documentation.txt
[19] https://github.com/github/docs/blob/main/CONTRIBUTING.md
[20] https://github.com/github/docs/blob/main/CONTRIBUTING.md
[21] https://docs.gitlab.com/ee/development/documentation/
[22] https://developer.chrome.com/docs/handbook/
[23] https://github.com/GoogleChrome/developer.chrome.com/tree/main/site/en/docs/handbook
[24] https://kubernetes.io/docs/contribute/
[25] https://docs.microsoft.com/contribute
[26] https://github.com/MicrosoftDocs/Contribute
[27] https://docs.newrelic.com/docs/style-guide/writing-docs/processes-procedures/create-edit-content/
[28] https://github.com/newrelic/docs-website/tree/develop/src/content/docs/style-guide

Example online style guides

A style guide provides authors with the best ways to write consistently and with the voice and tone that the organization wants for its brand. The Microsoft and Google style guides even have automated rulesets for the Vale prose linter, which you can read about in the section about testing docs.

- Apple Style Guide[29]
- Content style guide for GitHub Docs[30]
- Digital Ocean Technical Writing Guidelines[31]
- Google developer documentation style guide[32]
- Microsoft Writing Style Guide[33]
- Salesforce Style Guide for Documentation and User Interface Text[34]
- SUSE Documentation Style Guide[35]

Guiding contributors and authors

When you treat docs as code, your contributors must be able to author content smoothly and efficiently. To encourage more contributions, provide an easy-to-use authoring tool with a side-by-side comparison feature that closely matches the real output when possible. You also want to get people started with a great contributor's guide, enable both local and staged builds, and make sure that the deployed docs adhere to a style guide consistently.

---

[29] https://help.apple.com/applestyleguide/
[30] https://github.com/github/docs/blob/main/contributing/content-style-guide.md
[31] https://www.digitalocean.com/community/tutorials/digitalocean-s-technical-writing-guidelines
[32] https://developers.google.com/style
[33] https://docs.microsoft.com/en-us/style-guide/welcome/
[34] https://www.digitalocean.com/community/tutorials/digitalocean-s-technical-writing-guidelines
[35] https://github.com/SUSE/doc-styleguide

Automate builds so you can focus on writing

End the tedium, enable the robots.

Recite this mantra when you choose CI/CD tools.

- Continuous integration (CI) means that code is continuously tested, integrated with other code changes, and merged.
- Continuous deployment (CD) means that code is continuously deployed with each patch to the entire code base.

For docs, CI/CD means that content is continuously tested, merged with each patch, and deployed. Deploying, in this context, means publishing; for example, output files are copied to a web server for all to see.

If you already automate code builds and deployment, you can automate your docs builds and deployment.

Build and deployment tools

When you have simple markup and flat files for your docs source, you need to build those files into HTML that's connected to CSS and JavaScript, so that you can deploy the files to the web server. When treating docs as code, *build* means to parse and render to another format, and *deploy* means copy the files to the correct location on a configured server. When making an EPUB, for example, you build HTML and compress the files with some additional descriptive files into a single artifact. The build steps can differ depending on the deliverable. The deploy steps can change based on the method of delivery. For example, you could build translation artifacts and then deploy those to a translation server.

To build this book, we used these tools:

Tool	Used to
Visual Studio Code and Vim	Edit Markdown files.
GitHub and Git	Store files and perform version control.
Dockerfile	Automate the draft website and EPUB file builds.
GitHub Pages	Serve the draft website.

To make the docslikecode.com site, Anne used these tools:

Tool	Used to
Visual Studio Code and Vim	Edit Markdown files.
GitHub	Store files and perform version control.
So Simple	Provide a nice, responsive web theme.
Jekyll	Create HTML.
Netlify	Automate previews from a build branch on each Pull Request. Build the production website from the main branch.

CI/CD for docs

A docs build is typically triggered when updates are merged into the source code in the correct branch of a repo. A build usually consists of a bash script at the basic level, run by a deployment system such as GitHub Actions, Jenkins, Travis, or Circle CI. GitHub has many CI systems to choose from. You can learn more about them on the GitHub website in the Marketplace > Continuous integration category[36].

Benefits of using CI/CD for docs

When a fast-moving software team has multiple projects merging multiple changes every day, the docs system needs to keep up with that many changes. CI/CD enables that, so it's not only a benefit but a requirement in that type of environment.

CI/CD also means that reviewers don't have to build docs in their local environments. When the system builds a review draft, casual contributors and reviewers can avoid the overhead of downloading the patch, replicating the web server environment, and then building the docs. Instead, they can quickly see how a change is rendered when it is published. Because of the automation in the system, reviewers can review both the source and the output.

---

[36] https://github.com/marketplace?category=continuous-integration&query=&type=&verification=

The speed of the builds increases because writers can work on multiple patches at once while the cloud-based CI/CD continues to run.

With docs as code, using the same workflow that development and infrastructure teams use makes it easy for developers to contribute to docs.

Avoiding automation risks and pitfalls

Considering that writing is both a technical venture and an artistic endeavor, you have to strike a balance between what should be automated and what requires a manual set of steps. Some risks of not striking this balance could be publishing too soon or publishing incomplete docs.

To mitigate these risks teams should build trust among reviewers. Ideally the docs team writes a review guide and trains reviewers about using good judgment when reviewing patches. As long as reviewers have guidelines such as "It's better than what we have now," or "I've tested this and it works," or "This doc fix matches the reported bug I've investigated," the risks associated with publishing doc updates 50 to 100 times a day goes down. Automating the build and deployment processes gives reviewers confidence that the docs always build correctly and that the entire docs site is stable.

Some manuals are going to be release-independent, and some document the current release. For manuals that document a specific release, like an installation guide, the docs team can update the files for each new release and works in branches. The released document is updated with critical changes, and the document for the next release gets the majority of changes and gets published in a hidden place for easy review of the current draft. A docs team core reviewer can publish the release-specific guides publicly after the software is released, and then contributors start working on updates for the next release in new branches.

GitHub Pages hosted websites

GitHub offers a directly hosted solution for documentation called GitHub Pages[37]. This solution gives you a simple but powerful way to edit, push, and view live web pages. GitLab, Cloudflare, and BitBucket offer similar page hosting services.

You can register a domain name, create a repo, write a handful of pages, and then use GitHub Pages to publish to the domain name by changing a few settings and then pushing to the branch you indicate in those settings, such as the `gh-pages` branch.

Because you can map a domain name to the entire site, or simply use the provided `{githubusername}.github.io` sub-domains, you get nice web experiences branded for your project, organization, or company. The tutorial on the GitHub Pages[38] site offers a quick way to try it out, and the tutorial in this book shows how to use a specific Jekyll theme to build a website by using GitHub Pages.

The basic premise is that when you push changes to the `gh-pages` branch, you initiate a build that becomes web pages. You can even delete the default (or `main`) branch for a docs-only repo and always edit and push to the `gh-pages` branch.

If you use Jekyll with GitHub Pages, only certain Ruby Gems are supported to ensure the security of the hosted sites. Ruby Gems are packages that provide additional features, such as link checking. However, you can copy the output HTML files and related CSS and JavaScript files to the root of the repo and push those files to the `gh-pages` branch. That method uses GitHub Pages to host a static site, which lets you use any Ruby Gems or other Gem-packaged theme that you want.

You can also use Sphinx with GitHub Pages by adding a file named `.nojekyll` at the root of the repository. That indicates to GitHub that the files should simply be built and copied to the host. You can set up the repository to build to the `/docs` directory on the `main` branch, so that you are sure to keep the source and output HTML separated. You could also use a `_build` directory in the root of your repo as the served HTML.

---

[37] https://pages.github.com/
[38] https://pages.github.com/

GitHub Pages with Jekyll provides the following features:

- Automatically create site maps and an Atom feed when the theme provides them.

- Optimize the site for search engines, social media sharing, and redirects.

- Ensure that common repo metadata is available to the site.

- Enable the use of `@Mentions` of GitHub user names on sites built with GitHub Pages.

This tool is a great enabler of fast publishing from any repo. GitHub Pages has created some nice efficiencies for hosting documentation from GitHub. Here are some additional tips:

- Be sure to note the specifics for repo or organization names when using a custom domain name, which is explained clearly in the GitHub docs on custom domains.

- Use the Settings tab for the repo to indicate whether to publish the `main` branch or `gh-pages` branch, or use a `/docs` directory on the `main` branch for GitHub Pages.

- If you use a repo only for a website, you can delete the `main` branch and maintain only the `gh-pages` branch, as long as your site always uses only the Jekyll plug-ins that the GitHub Pages Gem supports. This simplification makes the purpose of the repo clear.

- When using a custom domain, you also create a `CNAME` file that indicates the domain name to GitHub Pages. Domain name registration costs about $10 per year, and you must see your domain registrar documenation to set up the domain name. If your DNS provider does not support `ALIAS` records on the root apex (@), you must create `A` records that point to IP addresses provided by GitHub, referring to their documentation.

Programming language considerations

The two languages and docs frameworks often used to build docs as code are Python with Sphinx, and Ruby with Jekyll.

When you learn Python, you realize how much readability, clarity, concise style, formatting, and opinion matter in this language. Whitespace is actually required and meaningful, unlike in other programming languages. Python is a dynamic programming language, and because it's interpreted, it often takes fewer lines of code to accomplish a task than in other, statically typed languages, such as Java and C++.

Ruby provides multiple ways to perform a task, unlike Python with its *one way is the most understandable way* philosophy. Ruby is dynamic, reflective, and quite general-purpose, and it predates Python by about four years.

You can do just about anything with either language. Both have excellent beginning frameworks for docs and websites. Both have a huge set of libraries for reuse. Both have a specialized tool chain to make lovely websites, and both offer strong community support and maintenance systems.

The maturity and wide adoption of both languages give writers a choice. To choose, consider the developer expertise that you already have in case you must extend the framework, such as by writing Sphinx extensions or Jekyll gems.

You might want to use a specific theme or organizational structure as you examine the web output. For the source, consider whether semantic markup is meaningful to your contributors and in your output. For semantic markup, reStructuredText (Python and Sphinx) has better support than Markdown (Ruby and Jekyll). Markdown has a history with multiple interpretations for output, but now has a specification for GitHub Flavored Markdown[39] (GFM).

Python: reStructuredText (RST) and Sphinx

The source for docs built with Sphinx in the Python community is called reStructuredText[40] (RST). It has an advantage over other simple markup languages because it provides semantic (meaningful) names to indicate whether inline markup is for a command or parameter element, for example.

One amazing aspect of the Sphinx and Python community is the Read the Docs[41] site, a community-maintained CI/CD system for documentation that supports both RST and Markdown as source

---

[39] https://github.github.com/gfm/
[40] https://www.sphinx-doc.org/en/stable/rest.html
[41] https://readthedocs.org/

formats. Started by Eric Holscher and other developers, the site hosts documentation for multiple projects, providing searchable and findable documentation pages.

Ruby: Markdown and Jekyll

Source files for docs in the Ruby ecosystem are Markdown. GitHub has its interpretation of Markdown, so any document files written in Markdown are easy to read on GitHub itself.

> **Query**: What if my source for documentation isn't Markdown or RST?
>
> If your content can be served from a web server or uploaded as PDF or eBook-formatted files, you can deploy in an automated fashion and use a docs-as-code system for documentation authoring and deploying.
>
> For example, let's say your favorite markup source is ASCIIDoc. There's a great write-up about how to use GitHub Pages for Pelican[42] with tips for helper tools to update the `gh-pages` branch continuously. Antora is a favorite static site generator with ASCIIDoc. Adding the `.nojekyll` file in the root of the repository enables you to publish with GitHub Pages using other source files and static site generators.

Web development tools

To do the web development necessary to make great interactions on websites, you must layer in more languages and markup with these interpreted languages. The list includes:

- JavaScript
- CSS
- Less or Sass, which compiles the CSS using variables
- Node.js
- Dependency organizers and library installers

You don't need to have a complete understanding of every framework, but be sure you have a decent theme that provides a nice user experience.

---

[42] https://docs.getpelican.com/en/latest/tips.html

Also, the Pandoc[43] command-line tool is useful when you must convert from one source file to another, especially in bulk or when you must automate.

Container images for docs

Automated builds might be the best part of the docs-as-code approach. Automation means that your memory-intensive builds won't cause your old laptop fan to spin out of control or your computer to die a swift memory-error death.

In the beginning of building docs at Rackspace, the team had a system of button clicks for publishing, with a server built with the Apache jclouds Java SDK, and running a Maven plug-in. When Rackspace teams enabled more contributors, the doc-build automation system changed from a home-grown and quirky publishing system to containers.

Although containers are not new, they are recently improved for use by developers. Containers let you serve an already installed and running software stack, such as Ruby with Jekyll.

You can use a Docker™ image for docs to build a Jekyll site locally. With Docker, you launch a local virtual machine (VM) that runs a container already created with Ruby and Jekyll to build the site in the VM. The process is straightforward: install Docker, get a prebuilt container image, run a command, and test your site through a URL that is served from a local container. Find an example Dockerfile and script in https://github.com/justwriteclick/versions-jekyll.

Additional uses of CI

You can also use continuous integration (CI) to integrate with a translation server for docs. Whenever a change merges, the CI infrastructure server uploads the current text automatically to the translation server, so that translators can directly translate and always have the latest strings. Once a day, a so-called *periodic* job is run on the CI infrastructure to download all translated strings from the translation servers to the docs repos, and a change is proposed with any new strings. The documentation

---

[43] https://pandoc.org/

team has a chance to run the translation import through the CI infrastructure together with a manual review.

Additionally, you can use the CI infrastructure to synchronize any shared files from one repo to a few others. For example, multiple repos can share a single glossary, together with translations of that glossary, thereby gaining efficiencies in reuse and preventing translation rework.

This type of configuration enables humans to run the test scripts for a final review.

We hope that this section gives you ideas about how you can use CI/CD for docs processes, including shared content and translated content. We have found the benefits far outweigh any risks in the approach. Our need to match with other teams means we adopted the continuous mentality to shipping early and often. Look at your docs, open source or otherwise, with an eye towards automation, and see what obvious solutions appear.

Thanks and credit to Andreas Jaeger

Andreas Jaeger and Anne Gentle wrote the CI/CD sections collaboratively. Andreas is involved with the OpenStack CI Infrastructure. He has contributed to various open-source projects for over 20 years and works for SUSE.

Test the docs: linting, inclusive language, and DocOps

Any time you talk about docs-as-code methods you eventually run into the term "DocOps"—a play on the term "DevOps." DevOps combines Development with Operations into one set of actions and skills that meet business needs without judging whether the task or role should be assigned to a developer or an operator. Instead, DevOps teams and DevOps tools look for efficiencies in automation, scalability, observability, and security while making software.

Teams work in tight iterations to release solutions as often as possible, which is possible and even expected thanks to lots of continuous testing and deployment, along with seeking input from stakeholders frequently. Naturally, DocOps means "DevOps for Docs," meaning combining your docs-as-code specialization work with frequent release activities as DevOps specialists would work on improving agile software releases.

You can read more about DocOps and contribute to the meaning on the Write the Docs site > Documentation Guide > DocOps page[44]. You don't have to use docs-as-code techniques to practice DocOps, but product versioning strategies, build scripts, and tools support may well be part of a good DocOps practice.

Treating docs as code goes hand-in-hand with automation because continuous integration involves plenty of testing and deploying. But don't set aside operations and automation only for the docs site itself. People are also doing clever work with automating jobs that help with triaging bugs, going through comments, managing the docs-as-code-related projects, translation work, and more. It's amazing what automation can do to free up your time for writing, coding examples, planning and thinking strategically, and constantly learning about new technology.

Automate project, release, and issue management

Teams have found ways to make reviews and releases more efficient by automatically triggering actions. You can base the trigger and actions on comments in pull requests, on the lack of responses over some time in an issue, on the GitHub username of the person (such as looking up employee IDs), or on where your docs or product is in the release cycle (such as a freeze period for translation work).

The automation helps with scale, security, and efficiency when you are working in public, external repositories that mirror private, internal repositories. For example, perhaps some files should not be changed by the external contributors for security or release freeze reasons (such as during critical translation point-in-time releases). Certainly, you can write up the explanations for those files remaining untouched in your external Contributor's

---

44 https://www.writethedocs.org/guide/doc-ops/

Guide, but also automating with friendly messages makes the work more efficient for everyone.

Triaging new issues for urgency, severity, and category becomes an important part of any documentation project. You can also use triaged issues with specific labels to prioritize or assign work amongst collaborators.

GitHub has project boards that use issue labels to move work along a path and to notify a group of maintainers or contributors when there's a new issue ready for work, and perhaps it's "low hanging fruit," meaning one that someone could pick up easily to complete quickly. Or perhaps the issue belongs in another doc repo entirely, or in a code examples repo, or maybe the issue is a customer support question. Using labels automatically helps categorize those or move them into a more appropriate location where another team can handle the person's question or need.

Automate testing

When you treat docs as code, you can make incremental changes over time that ensure high-quality and technically accurate docs. You can also follow practices that will help to ensure the quality of the content. Automated testing, for example, is valuable, although it can be difficult. You must also find ways to incorporate editing best practices in your quality controls and provide valuable insights. Don't risk alienating contributors by focusing on minor errors rather than valuing the contributor's technical contribution.

Which tests should you run? What value do these tests provide?

Of course, you want to ensure that the docs build. With a build tool such as Jekyll (Ruby) or Sphinx (Python), you can incorporate other automated tests.

You want to be careful when introducing tests on pull requests, as you do not want to "scare off" contributors by making it too difficult to pass tests. Make sure that your contributing guide explains all the tests and describes which ones are required or optional.

In a Gerrit system, automated tests are either voting or non-voting tests.

- **Voting tests**. These tests block a patch from going through the gate and merging. Voting tests save humans the time of downloading a patch locally and running tests manually.

- **Non-voting tests**. These tests report but do not block a patch from going through the gate. For example, *Are the translations still working with this patch?*

In GitHub, repositories can be set up for automated tests called Checks. Note that the prerequisites for setting up Checks from scratch are quite technical; Ruby programming skills, the Smee webhook payload delivery service, the `Octokit.rb` Ruby library for the GitHub REST API, and the Sinatra web framework are all prerequisites to create a Checks API CI server app[45]. The shortcut is to use CI solutions from the GitHub Marketplace.

For a GitHub pull request, you can block merges based on reviewers and you can also block pull requests based on whether automated tests pass.

- **Required Checks**. Any Checks in the CI server must be run against the commit.

- **Skipped Checks**. The commit can be reviewed without having the Checks run.

- **Successful Checks**. Indicates that a required Check ran successfully.

For docs-as-code systems, we recommend that you work towards your builds running these automated tests:

Voting	
File syntax	Is the language syntax correct in all files in the patch? This test checks the syntax of individual files and helps contributors quickly locate syntax errors.
Docs build	Do the docs in the patch build? This test checks the status of docs builds. Optionally, you can upload the built docs to a draft server so that reviewers can easily review the newly generated content to see how a change looks in HTML and PDF.
Links	Do all links work? This test checks whether links to external websites work. You can also mark this test as non-voting to accommodate the possibility of offline external websites.
Deleted files	Does the patch delete any files that are used by other deliverables? This test checks that all docs can still build if one file is removed.

---

[45] https://docs.github.com/en/developers/apps/guides/creating-ci-tests-with-the-checks-api

Non-voting	
Translated docs build	Do the translated docs in the patch build? This test checks the status of translated docs built through the toolchain.
Readability and scanning consistency	Is the content in the files consistent for easier reviews and readability? This test checks for extraneous whitespace, overly long lines, some unwanted Unicode characters, or proper formatting (JSON). Extra whitespace and overly long lines make it more difficult to review source files side-by-side. Instead of looking at a lot of red or green lines that identify white spaces, you can focus your reviews on content changes.

For additional efficiency, run these tests only if they are relevant to the patch. For example, if the patch does not delete any files, do not run the deleted files test.

> **Tip**: Examples of automation checks
>
> The Microsoft PowerShell repository runs automated checks on pull requests, and sometimes a machine spots a problem before a human can. For example, what's wrong with this set of tags? `<kbd>RightArrow</kdb>` Automated testing found it right away: `kbd` doesn't match `kdb`.

Another tests checks if the pull request exceeds a certain number of files. If so, the contributor is asked to make it smaller, which makes the work easier for the human reviewer.

Automated tests should help the human reviewer by taking away tedious tasks or checking for comparisons that are difficult for humans to make consistently.

There are plenty of discussions about automating spell checks, but such checks require a human for judgment calls or a large trusted dictionary for inclusions and exceptions. Refer to the section about prose linters for more information about content checks.

Prioritize technical reviews

The docs-as-code ecosystem values technical accuracy above all else and relies on reviewers to guarantee it.

It's hard to predict how long it takes to verify the technical accuracy in a docs patch. However, a human must double-check the technical accuracy of any docs contribution. To save time, you can set up environments to test user actions as part of a review. Make sure to include instructions for accessing test environments in your contributor's guide.

Check for style guidance, grammar, and spelling

In the 1970s at Bell Labs, a C developer created a tool named "lint" to check code for problems like extra unwanted lines or incorrect syntax, similar to trapping extra fluff in a clothes dryer. A category of tools called "linters" descended from that tool. These tools help avoid any extra discussion about code style and aesthetic choices during reviews, and a prose linter can do the same for doc reviews where a style guide has already been agreed upon.

Teams can use prose linters like Vale[46] to test their doc pull requests. Vale is open source and requires a ruleset. You can write one for your docs or use one already written such as the Microsoft writing style guide or Google developer style guides. An example rule is, "Don't use end punctuation in headings."

- Google developer documentation style guide[47] encoded with https://github.com/errata-ai/Google

- Microsoft Writing Style Guide[48] encoded with https://github.com/errata-ai/Microsoft

As alternatives or additions to the Vale checker, Grammarly and Acrolinx are tools with large AI-based natural language processing. Natural language processing is what makes these tools particularly helpful. For example, Acrolinx can differentiate

---

[46] https://github.com/errata-ai/vale
[47] https://developers.google.com/style
[48] https://docs.microsoft.com/en-us/style-guide/welcome/

that a "master's degree" is allowed when "master server" in an architecture description is not.

Acrolinx can be configured to run against each pull request and give a scorecard so that you can run a baseline and see how much improvement you can get on your doc sets. Acrolinx also encodes your company style guide in a set of comprehensive rules and policies that you share across all the repositories.

Inclusive language checks

The technical documentation world has long understood the power of words. We choose words carefully to convey the correct meaning and avoid harmful subtext. Many companies and organizations have taken up the responsibility as well to clear out biased choices of the past. We have the opportunity to modernize code and configuration text to replace terms like master/slave and blacklist/whitelist when they remind people of their hurtful historical meanings.

As an example, in 2020 GitHub moved from using `master` as the default branch name for repositories to using `main` going forward. They announced it in this blog post, The default branch for newly-created repositories is now main[49] and pointed to a repository, https://github.com/github/renaming/, where they could take in comments and input from the community.

Another example is the Microsoft style guide, which includes a section on Bias-free communication[50]. It provides a great reference and starting point if you want to start somewhere.

Fortunately, when you treat docs as code you have tools at hand that enable you to eliminate the use of these words. Plus you can enforce policies automatically on any changes to the documentation or code examples. One way to train writers locally is with a Virtual Studio Code extension that uses the "alex" linter which catches insensitive, inconsiderate writing. You can find the project and rulesets at: https://github.com/errata-ai/alex. Then, you could also run that linter's rules on pull requests to the repository. These tests could be for information only and

---

[49] https://github.blog/changelog/2020-10-01-the-default-branch-for-newly-created-repositories-is-now-main/

[50] https://docs.microsoft.com/en-us/style-guide/bias-free-communication

would not need to block or gate merging. Sometimes the goal for inclusive language is awareness and discussion before full policy enforcement.

Checking screenshots against the current interface

A great opportunity for web interface products is a chance to automatically capture and compare screenshots in the docs with the current product. Screenshot maintenance headaches, be gone! Automating screenshots would help many teams who want to give visual learners some clues about the product interface. But if it changes often needs real data to be loaded, it can be difficult to keep up and take realistic screenshots.

Docs as code techniques by themselves cannot solve the problems inherent to maintaining screenshots in documentation, but image automation tools work well in a docs-as-code CI pipeline. For example, Diffy, at https://diffy.website/, offers web UI automation testing. It can log in for you, let you ignore certain areas of the screen, and monitor the production environment, looking for changes.

Putting image captures into your CICD pipeline requires lots of development lift, but the results can save you enough time that the initial investment pays for itself.

Opportunities abound for testing the docs

Checking the docs with simple starter tests like a site build is a great starting point. You can add complexity from there including link checking, missing files, syntax and formatting problems, and how well the format will help reviewers see the differences in two files.

More opportunities are presenting themselves in grammar and inclusive language checkers. Natural language processors have advanced features that differentiate subtle differences in language use and context.

By testing the docs automatically, you free up reviewers' and writers' time to work on more difficult review tasks and higher-level improvements to the docs.

Review your docs as code

Content review fits well with quality assurance goals. Learn about the tools to use to review docs, which are the same tools that are used to review code.

One aspect of docs reviews is running automatic quality tests. See the "Automate testing" section for more information about automation. You can use a docs review checklist to ensure that your reviews are thorough and consistent.

Review rules and branch protection can provide detailed workflows for internal and external collaboration.

This section covers the GitHub and Gerrit review systems. Both systems are integrated with Git version control.

GitHub reviews

In 2016, GitHub added a set of user interface workflows that enable you to review patches. You can still do line-by-line comments with the **Add single comment** button. What's additional is the ability to comment on the overall patch and either request further changes or approve the changes. There's also a setting that lets administrators on the repo require that all pull requests must be approved before merging. Read the GitHub documentation for reviewing changes in pull requests[51].

You can assign reviewers to a PR and also ignore reviews if needed. Get agreement from your team about what review steps are necessary for merging a pull request. You can also read more about permissions for GitHub[52] when merging or closing pull requests. User accounts and organization accounts have finely tuned permissions.

---

[51] https://docs.github.com/en/pull-requests/collaborating-with-pull-requests/reviewing-changes-in-pull-requests

[52] https://docs.github.com/en/get-started/learning-about-github/access-permissions-on-github

Protected branches

You can add more protection to certain branches when you want to require strict review levels. Some teams allow for automatic publishing after the document meets certain criteria such as being free of broken links and checked by a set of tests and a single reviewer. Other teams would allow "self-merging," which means that the author can merge any change they make. Here are some varying levels of protection and approval based on GitHub's settings, which are based on what various open source teams have determined helps them with their understanding of history and trust of approvals.

Branch status	Possible methods of protecting or merging
Unprotected branch	No protection; author self-merges
Branch drafted; prior to review	Status checks; author makes sure tests pass
Branch ready for review	Review protection; team or person must review and approve
Branch reviewed and approved	Multiple protection options

Protection options:

1. Status checks on the branch must pass.
2. The branch must be up to date with the main branch.
3. A contributor license agreement requires a signed commit.
4. Conversations must be resolved in the review.
5. History must be linear; meaning all commits are rebased (or squashed) to one commit before going into the main branch.
6. Admins cannot override other rules and merge anyway.

Link to bugs in docs patches

As a best practice when working with docs contributors, let reviewers know the reason for your patch by referring to the bug or issue that the patch fixes. Such a reference helps reviewers compare the fix to the issue and judge whether the fix solves the stated problem.

For example, if you put a GitHub Issue number in the commit message, a link automatically goes to the Issue from the Pull Request once it's on GitHub. This link between a doc bug (GitHub Issue) and the contribution (GitHub Pull Request) makes it easier to tell whether the patch resolves the bug. Reviewers can click the link for the bug and read the comments on it.

Ensure that contributors can tell whether a bug has been accepted. To accomplish this in GitHub, add labels to issues. Then, when you add an issue link to your pull request and the pull request is merged, the issue is automatically labeled `Closed`. Provide label descriptions in your contributor guide to help people use labels as intended.

Projects have done a lot of work with automation for labels and processes, with bots that do clean up based on timing and other criteria. This saves human reviewers and contributors time and effort. For example, a bot can add the label "triage" to indicate that an Issue needs to be looked at by a human before work begins on it. Once someone verifies it's a real doc bug and adds correct labels, work can begin.

Comment automation to trigger actions on pull requests

Another part of the review process is the "sign-off," meaning that the author approves of merging the changes and they have completed the changes they intend to. By setting up meaningful labels that trigger certain automation scripts, you can enable read-only contributors to indicate that they want their change to merge.

And as long as you have other test criteria in place, you can confidently enable comment automation where a comment in the pull request adds a label that causes an action. The Review and sign off[53] section of the Microsoft Contributor Guide describes how their comment automation works. Their comments are `#sign-off`, `#hold-off`, and `#please-close`, which assigns either `ready-to-merge` or `do-not-merge` labels to the pull request, or closes the pull request entirely.

---

53 https://docs.microsoft.com/en-us/contribute/how-to-write-workflows-major

Designate a review team for a directory or collection of files

With GitHub, you have the capability to designate a review team. This setup works well for a scenario where a docs directory should be completely controlled by the documentation team. With a CODEOWNERS[54] file configuration, the documentation team does not need to ask for other reviewers for small changes like a typo or grammar fix. When a team member's account name or team name is in the CODEOWNERS file, you can protect the branch or folder from merges from any account not found in the file.

A CODEOWNERS file can be stored in a directory or in the .github folder, where many configuration files are stored by default. You can protect a branch or multiple branches based on a naming pattern match using protected branches[55]

A straightforward example involves protecting the docs folder in the main branch, where the main branch is the branch used to publish the documentation. In this example, the org name is justwriteclick, and the username is annegentle. You would name the file CODEOWNERS and store it in a docs directory.

Example CODEOWNERS file from an org named justwriteclick:

`@justwriteclick/annegentle`

You can also use an email address instead of an account name. See an example with lots of use cases in the GitHub documentation. Example of a CODEOWNERS file[56]

By using the notation, /docs/, you can ensure ownership of any docs directory, whether at the root of the repository or in a

---

[54] https://docs.github.com/en/repositories/managing-your-repositorys-settings-and-features/customizing-your-repository/about-code-owners

[55] https://docs.github.com/en/repositories/configuring-branches-and-merges-in-your-repository/managing-protected-branches/about-protected-branches

[56] https://docs.github.com/en/repositories/managing-your-repositorys-settings-and-features/customizing-your-repository/about-code-owners#example-of-a-codeowners-file.

subdirectory, such as `api/docs`. If you want to protect only the root `docs` directory, use `docs/*` instead.

Example `CODEOWNERS` file with a GitHub username, `@annegentle`:

```
/docs/ @annegentle
```

Example `CODEOWNERS` file with an email address, `annegentle@justwriteclick.com`:

```
/docs/ annegentle@justwriteclick.com
```

Then, ensure the branch is protected using the repository's **Setting > Branches > Branch protection rules > Add protection rule** button. Enter `main` for the **Branch name pattern** and then select **Require a pull request before merging**. The settings expand so you can also choose **Require review from Code Owners**. Once you have the settings, click the **Create** button at the bottom of the page. More details are in the GitHub Docs[57].

To see an example of this setup, look at the Microsoft 365 community documentation repository.[58] The `CODEOWNERS` file contains a team name, `@microsoftdocs/officedocs-admin`, and those team members can review and merge the list of documents in the `CODEOWNERS` file. The documents contain configuration information as well as the `CODEOWNERS` file itself.

Example `CODEOWNERS` file from MicrosoftDocs:

```
docfx.json                            @microsoftdocs/officedocs-admin

.openpublishing.build.ps1             @microsoftdocs/officedocs-admin

.openpublishing.publish.config.json   @microsoftdocs/officedocs-admin

CODEOWNERS                            @microsoftdocs/officedocs-admin

.acrolinx-config.edn                  @microsoftdocs/officedocs-admin
```

---

[57] https://docs.github.com/en/repositories/configuring-branches-and-merges-in-your-repository/managing-protected-branches/about-protected-branches

[58] https://github.com/MicrosoftDocs/microsoft-365-community

Measure improvements

By having documentation metrics, you can look for areas to improve to meet certain goals, such as adding contributors or decreasing doc complaints or errors. One great feature of GitHub is that it provides metrics for contributions and issues. You can visualize and measure improvements in the documentation itself, or you can focus on improving processes that enable more contributors.

Write down review expectations

Let your reviewers know what you expect in a review. Do you want to release quickly with technically accurate docs? Or, do you want to publish the docs on a specific release schedule? Do code patches merge only when the docs are up to par with the code?

Document your review expectations in the contributor guide for both new contributors and seasoned reviewers.

A review checklist like this one available from a docslikecode.com newsletter[59] can be a helpful tool to encourage thorough and consistent reviews. Categories provide groupings of questions to consider while you review, and the answer to the first question could prevent further review. For example:

☐ Do I have the configuration in place to test the instructions?

☐ Have I run all the commands in the doc changes and matched the results?

Use your team's priorities, tests, and style guide to modify the following examples to create your team's review checklist:

☐ Does the document build without errors?

☐ Is the output formatted as expected?

☐ Are the headings correct for the style guidance and overall organization?

---

[59] https://gallery.mailchimp.com/3828f8d87d82289b96ff8fd19/files/docs_review_checklist.pdf?utm_source=Coming+Soon+-+Docs+Like+Code&utm_campaign=c64a03b0c4-AUTOMATION_Docs_Like_Code_2&utm_medium=email&utm_term=0_cc1d483d59-c64a03b0c4-433282065

☐ Will the audience understand the context and why this information matters to them?

☐ Has data been properly redacted if necessary for security reasons?

☐ Are screenshots accurate for the version?

In GitHub, the – `[ ]` markup displays a cleared check box when you view it online, and the – `[x]` markup displays a checkmark in the check box. Read Task lists in all markdown documents[60] on the GitHub blog for details.

Write expectations for review turnaround times

You should set expectations for review turnaround time, writing it in the contributor's or reviewer's guide. Describe how you triage issues and pull request reviews so that people can understand severity levels and urgency levels, similar to a service level agreement. You can also set "slowdown" times when the core reviewing team will be on vacation, for example.

Gerrit reviews

Gerrit is a web-based collaboration tool that enables teams to write code together, and it can be used in the same way to review doc patches written in a source file format. It integrates closely with Git.

Gerrit provides a configurable *dashboard* that enables you to review across multiple repos by using search keywords and status indicators. For example, you can search for all API doc changes across dozens of repos by searching for specific file extensions in specific folders. Consistency across repos is necessary so that the search can locate all the relevant files that you would want to review.

---

[60] https://github.com/blog/1825-task-lists-in-all-markdown-documents

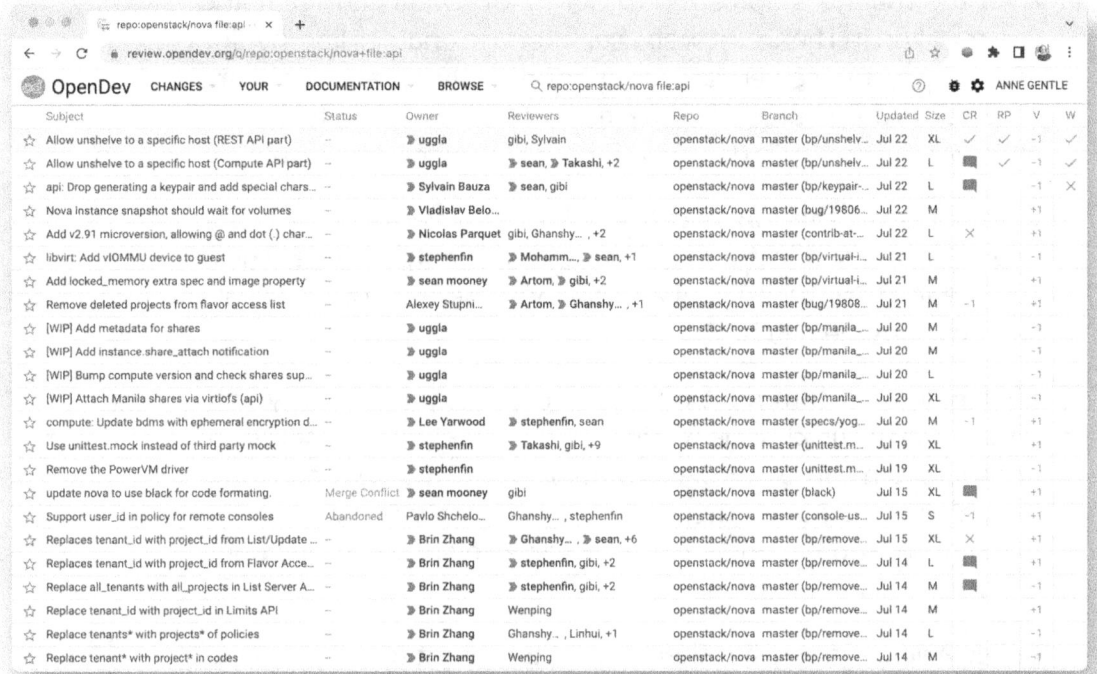

API reference documentation review dashboard

Gerrit also has a concept of *teams*, which can merge a change by voting.

OpenDev example: Gerrit reviews

To make changes to some open source code and docs repos, you use the Gerrit code review system. Gerrit is a web-based code collaboration and review tool. While pull requests provide the review mechanism in GitHub, Gerrit provides the review interface on top of any other source control system.

The basic workflow is that a docs contributor clones the docs repo, makes changes to the documents, tests them locally, commits them to Git, and then uploads them to the Gerrit instance. Gerrit then sends a change notification to the continuous integration (CI) services for software development.

After the notification from Gerrit arrives, the CI system runs the various tests that are configured for the repo.

After the change is uploaded to Gerrit, reviewers can see it and comment on it. The Gerrit web user interface enables line-by-line reviews. So, a reviewer can comment directly on any problems spotted in the source file. The tests also build the docs, so a reviewer can see built docs in HTML or PDF when applicable.

Comments on the patch can state what's wrong with it, ask questions for clarification, or state that the patch is fine. These comments help the original author and other reviewers update and evaluate the patch.

After reviewers comment on a patch, they can optionally vote for or against the change. Voting is an evaluation of whether or not the patch can be implemented. A positive vote means that the patch can be implemented; a negative vote means that the patch needs more work. A reviewer can also abstain from voting and provide only a comment.

All reviewers can register their vote in the Gerrit tool:

- 0. No score.

- +1. Looks good to me, but someone else must approve.

- −1. This patch needs further work before it can be merged.

A group of senior reviewers, or *core reviewers*, can also vote and approve a patch so that it can be published. Core reviewers can register the following votes:

- +2. Looks good to me (core reviewer).

- −2. Do not merge.

After two core reviewers vote +2, a core reviewer—normally the second one who gave a +2 to the patch—approves the patch, and then it is merged and published. A patch with negative review comments does not get approved, so the docs aren't published until consensus is reached with the proper approvals.

This screenshot shows a review for someone who is not a core reviewer, so only +1, 0, or -1 is available for voting purposes.

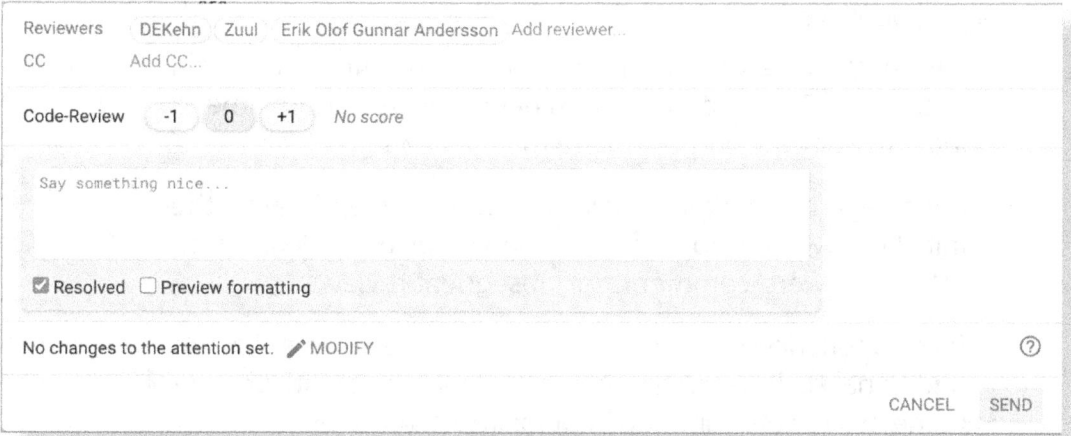

Voting system in Gerrit

The automatic testing also results in a vote during the review phase. When a change is approved, the system runs the tests again on the changes merged with the then current, updated Git repo to ensure that a merge did not introduce a failed build. A change can merge only if the test system reviews the change positively.

Coach better contributors

You can use reviews to coach your contributors' writing skills. Use comments in your contributors' pull requests to demonstrate good writing, suggest organizational changes, and instruct them on the finer (or the completely nonsensical) points of the English language.

Some people who speak English as a second language are fascinated with and enjoy writing in English. Others are shamed when they misuse English, so they are reluctant to write docs. The tech docs world is often English-centric, yet plenty of tech writers write first in French, German, or Japanese and then translate their content to English.

Writers might hesitate to work with others on deliverables, and developers might feel they don't have much to contribute to the docs. But in the docs-as-code world, no one person knows everything, and you can achieve better docs through collaborative writing and reviewing.

Some guidelines:

- Let newer reviewers do the reviews on the first day that a patch is up. This approach gives new reviewers more time to learn the doc set and complete their reviews.

- Encourage more experienced reviewers to step in after the initial review. This approach lets newer reviewers learn from other reviewers' comments in subsequent reviews.

- Pair experienced reviewers with newer reviewers. To tag additional reviewers, use the @ symbol with a GitHub user name in a review comment. For example, `@annegentle`.

- For small technical suggestions, patch the patch yourself. Never merge a patch, however, until the original author has a chance to review your changes.

- If your review might sound a bit harsh to a new reviewer, reach out on chat, IRC, or in a private email. Let contributors know that you aim to instruct and that you appreciate their work.

> **Tip:** The review balancing act
>
> At an internal developer conference in 2015, Anne wanted to know what other teams do about review quality. Specifically, how to improve the review experience from both sides. For example, she disliked marking only trivial mistakes or stylistic differences to the conventions.
>
> Everyone would rather do technical reviews and ensure consistency and quality, of course, but with time constraints, Anne had to find out how to get better at prioritizing what she reviewed.
>
> At this developer conference, people suggested a short waiting period. Yes. Wait to review and let others review first. Also, they suggested that contributors should patch other people's patches. Ensure that the original author still agrees, but make the docs experience easier and even enjoyable.

Create and fix doc issues

To notify others about doc bugs, you can create issues. For example, if you spot a typo in the docs, you can create an issue. After the docs team verifies that the issue is a true issue, you or someone else can submit a pull request to fix it.

For example, at the end of the calendar year, you might update the copyright year in the docs in response to a doc issue.

If the docs team uses labels effectively, you can click a label to find all related issues. For example, the Bootstrap docs repo uses a docs[61] label.

To find content to update, look for unassigned issues in the repo. If you can assign the issue to yourself, do it! Otherwise, comment on the issue to request that someone assign the issue to you. Then, begin work on the issue.

For large projects, the usual workflow is to fork the repo and issue a pull request. To find out how to make updates, look for a contributor guide or a `CONTRIBUTING` file in the repo. Ideally, the contributor guide tells you how to make updates. For example, see Contributing to Bootstrap[62].

For excellent examples of GitHub Pages, see GitHub Pages examples[63]. Peruse these repos to find existing source docs to update.

Deliver copy edits

If you have worked with a tech editor, you know that copy edits can be numerous and complex. Back when editors used a red pen to mark up printed-out pages, writers complained that editors made their content bleed!

Traditionally, the editor provides markup to a writer who updates the source files. However, when you work on GitHub with developer contributors, this method has limits. GitHub doesn't have a built-in way for editors to mark up changes in a file. And

---

[61] https://github.com/twbs/bootstrap/
issues?q=is%3Aissue+is%3Aopen+label%3Adocs

[62] https://github.com/twbs/bootstrap/blob/master/CONTRIBUTING.md

[63] https://github.com/showcases/github-pages-examples

developers might not be willing to or have the time to make tons of editorial changes.

Writers and editors can use a couple of techniques to deliver copy edits to other writers and developers. The technique that you use depends on whether contributors can incorporate editorial feedback and on the number of comments.

- If contributors can incorporate edits and you have a small number of comments, enter line comments in a pull request. For numerous comments, copy the content to an app such as Microsoft Word or Google Docs and then mark it up. Word has easy-to-use track changes and commenting features that enable an editor to show changed content, ask questions, and make comments. Google Docs has similar features. Then, create a PDF from the marked-up doc and attach that PDF to the pull request or a separate issue. The original author can then make updates in one or more pull requests, or the team can triage the issue to verify the editor's suggested changes.

- If contributors can't incorporate edits, you can update the file itself and create a pull request with your changes. When you create the pull request, you can enter line comments with additional questions or suggestions. The contributor can compare the differences between the original and changed content and decide whether to merge the pull request as-is or make updates before merging.

When contributors make updates, they can see their mistakes more clearly, learn about style and writing guidelines, and become better contributors. However, workloads might not permit contributors to take the extra time to make updates. So the key is flexibility.

> **Tip**: Professional technical publishing of books on GitHub
>
> Technical publisher O'Reilly has used Git repos for years when working with technical authors on content. For example, when working with the production team at O'Reilly on a line-by-line copy-edit of an entire Operations Guide, the team had multiple writers at the ready, able to enter their edits. The production team

emailed the exact process. Their quality control process had both a copy-edit stage and copy-edit review stage, where someone verified that all suggested edits were made. Here's an excerpt from their acceptance email when the manuscript went to production:

 The purpose of the copy edit is to check for errors in structure, language, grammar, spelling, and formatting. The copy editor reads the entire manuscript and marks any errors. The copy edit is sometimes done in batches (chapter by chapter), if appropriate.

Scaling large incoming review requests

Team managers found that the most helpful tools for scaling are: * Use automated checkers such as Vale as a prose linter and automatic link checkers. Those are a force multiplier to give the human reviewers confidence in the patch so they can focus on the more difficult parts of the review such as technical reviews. * Train reviewers on "level of review" so that they know whether to do a technical review only or also look at say, the structure or copyedits. * Ensure the goals for an article are clear so the reviewer can match the goals with the patch. * Make sure reviewers understand the overall structure of the site and how a page sits within the site. * Measure rates of reviews and determine explanations for dips in review turnaround times. Find solutions if problems are detectable.

As an example, the GitLab team tracks their merge rate as a Key Performance Indicator as shown in their publicly available performance indicator charts per department[64]. The merge rate is calculated as the number of merge requests divided by the number of team members. They have a goal of 55 per month but noted that in June and July of 2022 they were at 49 per month, likely due to multiple people taking time off, plus onboarding new hires. This indicator is set with 12 to 15 team members and 500-700 merge requests per month. They also have a tool that shows a dashboard with assignments by docs maintainer for

---

[64] https://about.gitlab.com/handbook/engineering/ux/performance-indicators/

the last 30 days[65] which allows a coordinator to determine an available reviewer.

Putting reviews together with reviewers

Reviews can be done partly by automatic tests, and many reviews are completed by people using review systems. Ensure that you understand the permissions for merging and closing pull requests and then look for ways to guide reviewers. Setting expectations for reviewers and linking to doc bugs in the commit messages help reviewers get better at reviewing documentation. A review checklist is also a great way to indicate top priorities and completeness to reviewers. You can use labels and comments to automate some actions on pull requests which helps when you need to review large numbers of changes.

Versions and releases: publish docs as code

Generally, you publish the product docs when developers release the software.

You could publish your docs more often than the developers release their code but when you treat docs as code, the source files for your docs often use the same version control system that the code uses.

You can exploit the features of this version control system for your docs. For example, when a reader views a back-level page, a banner can appear that links to the latest version of the docs.

The docs URL can also indicate the version of the docs, such as `/latest/` or `/4.1.3/`. Or, the URL can indicate the release status, such as `stable` or `beta`, or the docs branch, such as `main` or `develop`.

---

[65] https://gitlab-org.gitlab.io/gitlab-roulette/?sortKey=stats.avg7&order=-1&visible=maintainer%7Cdocs

Define a docs version

The Write the Docs[66] Sphinx theme is recognized for its excellent version badge design. On the site, a small version badge floats on each page. To choose another docs version or download another output format, the reader simply clicks an arrow control.

RTD Version Badge Expanded Example

To implement this version badge, you specify the version number in a parameter in a `conf.py` file with each build. Maintenance is straightforward and you can integrate this parameter with the version control system.

---

[66] https://www.writethedocs.org/

The Django documentation site[67] also implements a version badge.

- Prev: Soluções de Problemas
- Next: Aplicações
- Table   1.9   te   1.10   dev   Documentation version: **1.11**

Language: **pt-br**

Django Version Badge Example

You can also use the docs version parameter to:

- Show a banner that lets readers navigate to the latest or different versions of the docs.
- Use the URL to show the docs version, release status, and docs branch.

For example, a `/latest/stable/main/my-docs` URL indicates that the docs are a stable release of the latest docs from the `main` branch.

Other scenarios to consider:

An unsupported version is no longer available online

You can redirect requests for unsupported versions to a newer version. However, that strategy sometimes results in outdated content. When a product stores source files in a public location, you may want to include instructions for how to get and build the docs for the unsupported version locally for personal use. Because your projects and organizations might have different legal requirements for unsupported doc requests, plan to accommodate those needs.

A page exists for one but not another release

For example, when the reader views a new page for the v1.2 release and then switches to an earlier version of the docs, the new page does not exist in the earlier version. To handle this situation, you can create a page that describes the reasons that the reader should upgrade to the v1.2 release.

---

[67] https://docs.djangoproject.com/

Your static site uses a single landing page to reflect multiple releases

Maybe the product manager wants to highlight a new feature or wants to show consistency release-over-release on the single landing page for the site. You must determine what to show on your landing page. Make sure that casual readers who search for a certain release find a landing page that addresses that release.

You must retract a version for security reasons

Do you simply redirect the reader to a later or earlier version? The URL to which you redirect the reader depends on which version your service uses instead of the retracted version.

A service skips a semantic version

What if your release skips from v4.1.1 to v4.1.4? Do you need to explain why v4.1.2 and v4.1.3 are missing? Use release notes to explain any oddities in release number order.

Define a docs locale

You can also use a configuration parameter to define when a translation is available for a particular version. When you indicate a docs version, you might also consider displaying the language, even if you do not have translated content available. It pairs nicely with versions and lets your design grow into further release models if needed. So think ahead about your URL containing `/de/latest/` or `/es/2.4.23/` and a language indicator on the page with the version indicator to let people know which version they are reading and if it's available in another language.

Notes on translations and continuous integration

Many open source projects provide tightly-coupled integration with translation teams from around the world. When the typical cycle for documentation is to freeze the English version so that the other language versions can be released as well, it's interesting to think about another language being the starting point. The automation that enables the system to have scripted imports and exports while enabling human translators rather than machines can be used in either direction.

While the docs usually start with written English, some open source communities write a document in their native language first and then provide translations to English. This model provides successful documentation to multiple users and emphasizes that English is not the only language for technical content.

Docs source version control

Will you use tags, branches, or releases to provide version control for your documentation source? When you choose one or even multiple methods, how do you use tags, branches, or releases to automate publication and provide a release of documentation? In some cases, you are tied to the publication methods, whether pushing to an external website that has a version-control system or producing flat files that must reside in a particular directory structure.

Realize that the complexity increases as you add requirements for protecting source access. When you also must protect output access, you have multiplied the complexity of your solution. While open source projects can work on the docs in the open while drafting, many software projects must write and review docs with a limited group of collaborators, such as in a private repo. This requirement may increase the cost of continuous integrations, for example. When you also must protect draft copies of published documentation, you need a staging server or a login system.

Each possible solution for the requirement increases in complexity as you share and reveal less and less during both development and production. Development costs also increase. Be ready to consider the trade-offs as you analyze.

Versions, tags, and branches with Git for docs source

When using Git, you can apply a tag to a particular commit by pointing to that point in the history of the repository. A light-weight tag is a pointer to a specific commit but isn't used for releases, as you likely want more information about a tag to release the docs. An annotated tag has the checksum, a label, and information about who did the tagging and on which date.

To tag a repository at a particular point in the history, refer to Git Basics - Tagging[68]. When you tag locally, you also push those tags so that other collaborators can see them when cloning or pulling from the repo.

Versions in GitHub are based on Git tags. You can use the GitHub web interface to create your repository's releases[69]. When you do so, you can download a compressed file of all the source files in the repo at that moment in time. You can base a version on any branch, and tags are used to indicate to contributors the version number of the doc source files. Because of the limits of pairing tags with a compressed file, and because tags must be re-pushed if changed, it's generally not a best practice to "move" a tag once it's already on the contributor's local environments.

For strict release environments, where a new release would be required if the tag changed, you may not want to set expectations of pairing tags with releases even though GitHub does. Generally, it's best to use the same system as the overarching software and choose the system with some familiarity for contributors or readers who tend to read doc source files rather than output content.

Use case: How Read the Docs supports versions

More than 80,000 projects use the Read the Docs infrastructure system to automate their documentation builds with publishing. Their versioning implementation[70] is based on the version control system and the default branch used there. For example, Git-based version control often uses `main` as the default branch, but you can change that label to the `develop` branch depending on your desired workflow.

---

[68] https://git-scm.com/book/en/v2/Git-Basics-Tagging
[69] https://help.github.com/articles/creating-releases/
[70] https://docs.readthedocs.io/en/latest/versions.html

Use case: How a software product doc team uses versions, branches, and tags

A docs team can use tags and stable branches to version and archive doc source files. One docs team uses tags on the common docs repo, marking a point in time for the docs that went with a particular release of the product. The team uses `stable/releasename` branches for the supported releases of the product. By using tags and branches, the team can mark a moment in time with a tag, but continue to work in a branch if needed. The team wrote a documented process for releasing the docs site in the README for the repo.

When someone needs to read the docs at a point in time for an unsupported release, they can get a copy of the tagged repository and build from those files.

When a contributor wants to continue to publish changes to a past release that is still supported, they back-port changes to the `stable/releasename` branch and publish the docs site again.

Both of these tasks, tagging, and branching for release, are well-documented so the release tasks are shared by team members. Ideally, the need for a cherry-pick and back-port happens rarely, and the team does not need to move the git commit value on a tag for an end-of-life release. Then, when someone must build docs from an end-of-life release, they can follow instructions to do so.

Releases for documentation sites

Because web-based documentation sites are often released once they are publicly available, you might need to pre-plan in stages to ensure the release of a docs site goes smoothly. Similar to an online service, you should start to think of your docs site in terms of testing, staging, and production. A testing area for doc builds helps you try out different designs, test and review patches to the docs, or try test processes and workflows. A staging area enables writers to see what their output will look like when published, as a staging area should be as identical to production as possible. You could consider a staging area to be like the backstage of a play. Only certain readers can have access to a staging area for a docs site.

Also, consider that your doc tools should be released and finalized before a final release of a documentation set. You should be able to replicate the exact output from the tool as the tool existed at the point in time of a release. These considerations may have different legal or contractual requirements in your organization.

Archives contain all files for a release

When you create a version in GitHub, you are likely creating an archive of the source files. If you also want to archive the built files, you must check the built files into a branch and create a version that contains the built files as an archive. Generally speaking, repos do not contain compiled or built files, so you must complete this additional configuration step to create an archive for docs.

Examples of releasing documentation

You can recognize two patterns that are well-established in the software projects world. One release pattern is a continuous publication pattern of the website, which is not coordinated with other projects, known as `independent`. An example might be User Guides, when not much changes during release-to-release of the products or services. Another is `cycle-with-milestones`, where the document needs to be published at pre-determined times to synchronize with a project or service's release. A Reference Guide could be an example of this doc release model, because new configuration options are introduced and only available at a certain released point of the software. It's important to ensure the page documenting those configuration options indicates in which release you can use that option. Also, that document must clearly show which options have been deprecated in a given release.

For a service that has new features every six weeks or so, you must provide release notes in addition to new pages that describe how to use the new features. A well-written branch process helps with these different needed release patterns.

When reading a page about a feature your service didn't have yet, you need a way to get to the page relevant to the product you have installed. Consider both the reader's experience and the contributor's experience when releasing a documentation site or other deliverable. Find an example written as a proof-of-concept using the static site generator, Jekyll, at https://github.com/justwriteclick/versions-jekyll.

This example demonstrates using version control for source and output, with output directories defining the version in the output. The HTML and CSS and JavaScript files can also work in concert to provide the user experience for selecting and reading another version, stored in the output directories.

Perspectives on versions: author and reader

Because of the different perspectives on both source and deliverables for documentation, you must think about versions, archives, and possibly translations when treating docs as code. Also consider the two perspectives on versions: the contributor's views and the reader's views of the version value. Each version implementation can differ slightly depending on the consumer of the version number. As you reveal content throughout the process, the amount of control you need over the source, the output, or both, increases the complexity of the solution.

Chapter 4

Lessons learned with docs as code

Lessons learned with docs as code

People wonder whether treating docs as code can work in their environment. Although outcomes matter and this movement aims to exceed users' expectations, the disruptions in your team's daily tasks, work expectations, and areas of control can make this transition difficult.

For example, think about what goes through employees' minds when they face a new system:

> "Can this scale?"

> "Can my team maintain the tooling required?"

> "What do you mean I have to review 10 patches before getting to write myself?"

> "I'm a programmer, not a writer; why do I have to add docs to my task list?"

> "Work was easier when I could send it to the tech writers to handle."

> "I am the owner of the configuration docs. Why would I let others work on something I've controlled for years? They might mess it up."

As this guide has shown, the shift to treating docs as code includes a complex overhaul of attitudes, processes, toolsets, and expectations. However, many teams work through these difficulties with great rewards. To ease the transition to using more docs-as-code techniques, look for these opportunities:

Find your community and learn from others

Many people have been treating docs as code for years, and the Write the Docs community[1] is the place to find other "documentarians" who appreciate this set of tools and techniques.

The Write the Docs conferences take place each year, and regional meet-up groups talk about great techniques for documentation. In fact, many presentations describe integrating documentation practices with code practices:

---

[1] https://www.writethedocs.org/

- 2022 North America conference: Marcia Riefer Johnston and Dave May[2] co-presented their experiences starting docs as code processes and tools at Amazon Web Services (AWS) and how they iterated and improved as they learned along the way.

- 2021 North America conference: Swapnil Ogale[3] talked about "Putting the"tech" in technical writer" and how docs as code tools and processes provide opportunities to prove a writer's technical skillsets.

- 2019 Europe conference: Jen Lambourne[4] presented "The UK government meets docs as code" where she described her team's experience at the UK Government Digital Service implementing docs as code using a "choose your own adventure" model.

- 2018 Europe conference: Predrag Mandic[5] described how to "Run your documentation" meaning test docs with the product including integrations, empower documentarians, and make sure the customer experience matches the docs.

- 2017 North America conference: Jodie Putrino[6] presented "Treating documentation like code: a practical account" to share her experiences at F5 Networks.

- 2016 North America conference: A panel[7] of folks from Rackspace, Microsoft, Balsamiq, and Twitter talked about how they are adopting these practices.

- 2016 Europe conference: Margaret Eker and Jennifer Roundeau[8], and Rachel Whitten[9] talked about docs-as-code workflows in practice.

- 2015 North America conference: Riona MacNamara[10] spoke about how adopting docs-as-code techniques has completely transformed how Google does its documentation.

---

[2] https://www.youtube.com/watch?v=Cxuo3udElcE
[3] https://www.youtube.com/watch?v=FQ7DkPOw3Cc
[4] https://www.youtube.com/watch?v=Ql9Il7tssik
[5] https://www.youtube.com/watch?v=oW7rWJ2xNZU
[6] https://www.youtube.com/watch?v=Mzu-c-FoOdw
[7] https://www.youtube.com/watch?v=Y2TGwUPb8R4
[8] https://www.youtube.com/watch?v=JvRd7MmAxPw
[9] https://www.youtube.com/watch?v=dHdBsNxtKel
[10] https://www.youtube.com/watch?v=EnB8GtPuauw

Create a great web experience

Your contributors want to know where the content goes. For them, communicate the layout of the site and use meaningful directory- and file-naming conventions.

For readers, provide excellent navigation systems and ways to find what they need.

Because traditional tech pubs toolsets aren't good at creating wonderful, responsive websites that provide an amazing docs experience, modernize your toolset. A natural tension exists between producing artisan, hand-crafted docs and producing tens of thousands of web pages that still offer a great experience.

Equip your contributors with a style guide

To build a trusted set of great reviewers, you must give them a style guide to use while writing or reviewing.

Unless you establish standards for terminology, your contributors end up arguing about which name lands in the source. For example, is it *plugin* or *plug-in*? A style guide frees your contributors to focus on what's most valuable: technical reviews. Instead of calling out grammatical errors or capitalization consistency issues, reviewers can point writers to the style guide.

A style guide will also be helpful when you create tools to automate quality checks. Refer to the authoring section for a list of online style guides.

Empower your contributors

When you engage a large number of contributors, many of whom you don't personally manage and direct, you can lose control of what gets written. You don't get to pick what is written first, second, or third, or even whether something is written at all.

Create a culture where contributors are empowered to prioritize work: let them determine what to work on first, second, and third. Be ready to trade control for many other benefits. And be sure to have processes and systems in place to triage docs issues.

The most organized and docs-ready open source projects and organizations can get funding for a technical writing project through the Google Season of Docs program[11]. A technical writer can look for opportunities in open source communities to help with docs and possibly gain experience with docs as code tooling.

Write a contributor's guide

Most code repos use a README file to explain how the contents work. When you are writing documentation in a source code repo, use the README file to explain exactly how someone can contribute, to display build status, and to describe any automation tools and how to use those locally to test changes.

You can also add a `CONTRIBUTING.md` file to the root of the repo to describe guidelines for repository contributors[12]. Describe what contributors can expect at review time, or how to give their patch its best chance in landing in the source, or how to log an issue. When a contributor uses the GitHub web UI to create an issue or pull request, they see a link to the `CONTRIBUTING.md` file in the web UI.

---

[11] https://developers.google.com/season-of-docs/

[12] https://docs.github.com/en/communities/setting-up-your-project-for-healthy-contributions/setting-guidelines-for-repository-contributors

You can also provide templates and examples of the best work you have ever seen, and point contributors to your style guide.

The 18F digital services agency at the United States General Services Administration provides the 18F Content Guide[13], which is a good example of a content guide.

> **Query**: Which comes first, the style guide or the contributor guide?
>
> Enough open, available style guides exist now that you can point to an existing style guide for starters, and focus on writing a contributor guide. For example, choose either the Google developer documentation style guide[14] or Microsoft Writing Style Guide[15], then write a basic contributing guide pointing to the style guide you like the best. Over time, you can write guidelines in your contributor guide for where your style may diverge from the original selected style guide.

Write a reviewer's guide

In addition to a contributor's guide, or as a section in your contributor's guide, write a reviewer's guide for the docs. A "good problem" to have is a large number of pull requests to review, but it can also be overwhelming. For example, for large projects with multiple repositories, some days there are over 100 changes that must be reviewed for the docs. This amount is intimidating, and the toughest part of the job can be disappointing people by not reviewing their docs changes in a timely manner. Write up a service level agreement for reviews to meet expectations for turnaround times within your team and the community. You can notify the community of times when the core team will be taking a break.

---

[13] https://content-guide.18f.gov/
[14] https://developers.google.com/style
[15] https://docs.microsoft.com/en-us/style-guide/welcome/

A reviewer's guide can contain a checklist for the docs and also explain any automated tests related to the docs pull requests. By providing a reviewer's guide you can recruit more reviewers, since those contributors provide as much or more value than the writers many times.

Build in continuous integration for docs

Automate builds and quality checks so that you spend your time delivering great content, not running builds. Let the robots perform the builds, and use tools that don't require a seat license so that the writers don't have to build from their computers only.

To succeed at automation, use tools that enable automation, like static site generators with a command line interface, content linters like Vale, or Markdown files that contain Bash scripts that automate an installation process for your product.

When you must make changes across multiple repositories and deliverables, be ready to learn Bash scripting or get help with it. Fortunately, text manipulation is pretty powerful at the command line. Unfortunately, some team members or outside contributors might be using non-Linux-based systems and have to find tools to help with this problem.

Teach everyone to respect the docs

Depending on your contributor base, you influence more people to respect the docs and encourage more contributors to improve the docs when the docs are in a code system.

Consider teaching developers to review code patches that also contain docs changes. Encourage them to merge code only after the docs are up to par with the code changes. Write a reviewer's guide within the contributor's guide to get better docs reviews.

Test and measure outcomes

Much of the techniques and technology are still in early stages of a software lifecycle, so do not cement yourself into particular workflows or tools. Be ready to change a process if it's simply not working well for the team. Test the outcomes in two ways: with web analytics for the success of the output, and with contributor analytics for the success of the source creation and maintenance.

For contribution patterns, you can use the GitHub UI to look for metrics. For an example, look at the Contributors graph[16] for the popular Microsoft Azure documentation. You can also use scripts from the Cloud Native Computing Foundation. They have done extensive work on data graphing for contributors with their Velocity project to analyze growth in participation in open source projects[17].

Look for a natural next step, using data queries to detect the best practices for docs-as-code techniques by analyzing contributor and reviewer data. Watch for opportunities to improve when you see best practices emerge based on data from large projects with historical data, or projects similar to your own. Team size, release frequency, commit frequency, ratios of docs and code metrics, and specialty areas related to docs and code collaboration are all interesting to study.

Set up a Git support chat room

Git can take a while to learn, and definitely takes hands-on practice. There's no shortcuts when learning Git. So each person may know a little more or less than another team member when it comes to Git. One team set up a special chat room just for Git help, so anyone could come to that room any time they

---

[16] https://github.com/MicrosoftDocs/azure-docs/graphs/contributors
[17] https://github.com/cncf/velocity

felt uncertain about what to do next with Git. Setting up a "Git Support" area for the team makes merge conflicts a little more bearable. Merge conflicts happen when Git tries to merge two changes but can't decide which change should be the final solution in the branch. When this happens, Git puts in markers like <<<<<<< HEAD and ======= so that a human can review which change should be the final change in the branch. The more you practice merge conflicts, the less "scary" they seem. And having a friendly team member ready to help makes Git less intimidating.

Summary

In some ways, this entire book is full of lessons learned. By stepping back and looking for best practices in code integration for docs, you can continuously improve your doc authoring, building, publishing, and consumption.

The main goal of treating docs as code is to fit in with the ecosystem, so that you can gain collaborators who know their stuff and keep the contribution tasks as simple as possible (but no simpler).

www.ingramcontent.com/pod-product-compliance
Lightning Source LLC
Chambersburg PA
CBHW081130170526
45165CB00008B/2625